Practical
RF Power Design
Techniques

Practical RF Power Design Techniques

Irving M. Gottlieb

TAB Books
Division of McGraw-Hill
New York San Francisco Washington, D.C. Auckland Bogotá
Caracas Lisbon London Madrid Mexico City Milan
Montreal New Delhi San Juan Singapore
Sydney Tokyo Toronto

© 1993 by **TAB Books**.
TAB Books is a division of McGraw-Hill, Inc.

Printed in the United States of America. All rights reserved. The publisher takes no responsibility for the use of any of the materials or methods described in this book, nor for the products thereof.

      6 7 8 9 0   BKM BKM   9 0 9

Library of Congress Cataloging-in-Publication Data

Gottlieb, Irving M.
   Practical RF power design techniques / by Irving M. Gottlieb.
     p. cm.
   Includes index.
   ISBN 0-8306-4130-0 (hard)   ISBN 0-8306-4129-7 (paper)
   1. Power electronics. 2. Electronic apparatus and appliances-
-Power supply. 3. Amplifiers, Radio frequency. I. Title.
TK7881.15.G72 1993
621.381′044—dc20   92-5599
                                                                 CIP

Acquisitions Editor: Roland S. Phelps
Book Editor: Andrew Yoder
Director of Production: Katherine G. Brown   EL1
Series Design: Jaclyn J. Boone   4228

Contents

Acknowledgments *ix*

Introduction *xi*

1 Transistors versus tubes *1*
What is radio-frequency power? *1*
Why should RF power merit special attention? *3*
Mechanism of radiation *3*
Why reconsider tubes? *5*
    Advantages of tubes for RF work *6*
    Advantages of solid-state RF power devices *6*
    Class-C tube RF amplifier *7*
    Linear RF amplifier *10*
    More-practical linear amplifiers *11*
    Transistor version of the grounded-grid amplifier *13*
    Solid-state approach to high power *13*
    Wilkinson power-combining technique *16*
    Amplifier module *18*
    Low-power RF amplifier module *20*
    Test circuits *23*
    Layout diagrams and PC-board art *23*
    Amplifier chains *24*
Low-frequency applications *26*
    Basic SCR inverter *27*
    Low-frequency solid-state transmitters *31*

The phenomenon of skin effect *35*
    A practical tip for winding toroidal inductors *39*
Simple RF power measurements and some simple pitfalls *42*

2   The bipolar transistor in RF power applications *47*

What is unique about RF power transistors? *47*
    Gain-leveling provisions *49*
    Basic transistor class-C amplifier *52*
Matching networks *53*
    Parallel-tuned output networks *55*
    Output circuits for medium- and high-power amplifiers *57*
Survival of transistors in RF service *58*
    Techniques to reduce hot-spotting in RF power transistors *59*
Basic problem of transistor RF power circuits *63*
    RF grounding of the common lead *65*
    Transistor frequency multipliers *65*
Selecting the RF transistor *67*
Avoiding a pitfall *69*

3   The field-effect transistor in RF power applications *75*

Junction field-effect transistor (JFET) *75*
    Simple circuitry at low-power Levels *77*
    Power FETs: "solid-state tubes" *79*
    V-groove structure of the power FETs *81*
    Some revealing features of the power MOSFET *82*
    Input circuit of the VMOS power FET *84*
    Output circuit of the VMOS power FET *87*
    Class-D RF amplifiers *89*
    A family of MOSFET giants *92*
    Three circuit configurations *94*
Disadvantages of the power MOSFET *95*
Avant-garde FETs and RF techniques *96*
The IGBT: superb performance at the low end of the RF spectrum *98*

4   Impedance-matching networks *101*

Basic considerations *101*
General design approach *102*
Basic mechanism of impedance transformation *103*
    Derivation of the basic Pi network *107*
    Derivation of basic tee network *108*
Wave purity in transistor amplifiers *108*
    Selection of network types for transistor RF amplifiers *109*
Some facts that pertain to all networks *110*
    Output impedance of bipolar transistors *111*
    Basic impedance-matching networks used with transistor amplifiers *112*

Computerized network solutions   *115*
Transistor impedance information   *125*
Scattering or sparameters   *128*
Impedance matching: circuits and semantics   *133*
    A closer look at the output matching situation   *134*
    Fixing the inductance of coupling, bypassing, and dc-blocking capacitors   *136*
    What to do about the inductance of network capacitors   *137*
Harmonic filters   *141*
    Design example of a 30-MHz harmonic filter   *143*
    High-pass versions of the L impedance-match network   *145*
    The two-control tee-circuit antenna tuner   *146*
    The SPC transmatch impedance matcher   *147*
    The Pi-L network   *148*
    A unique four-element impedance-transforming network   *150*

5 Applications of transmission-line elements to RF power circuitry   *153*

Use of transmission-line elements   *154*
Characteristic impedance of transmission lines   *154*
    ¼-λ lines   *155*
    ⅛-λ lines   *159*
    Other lines for impedance matching   *164*
    Transmission-line connections between driver and power amplifiers   *167*
    Special considerations when designing with stripline elements   *175*
    Reflectometer-type instruments   *178*
    Use of the SWR analyzer for quick and convenient RF measurements   *187*

6 Low-power applications   *191*

Microwave doppler radar system   *191*
1-MHz JFET crystal oscillator   *192*
JFET frequency doubler   *194*
Class-C amplifier: 1.5 W at 50 MHz   *195*
JFET broadband linear amplifier   *196*
Power MOSFET driver for a broadband linear amplifier system   *197*
Low-power 2-M linear power MOSFET amplifiers   *198*
135-MHz AM amplifier 200
    Doubling the output power (almost)   *202*
Complementary-symmetry driver amplifier for SSB   *204*
2-W 2.5-GHz oscillator that uses the common-collector circuit   *206*
Tapered-stripline microwave amplifier   *209*
Quadrature amplifier circuit   *212*

7 Medium- and high-power applications *215*

32-W marine-band amplifier *215*
70-W 2-M broadband linear amplifier *216*
80 W of class-C power *218*
Mobile 80-W 175-MHz FM amplifier *221*
Respectable power levels from MOSFET devices *224*
Implementation of the balanced transistor *227*
A 150-W 28-MHz test amplifier *227*
1 kW from 2 to 30 MHz *230*
    Thermal-tracking bias source *233*
    Input power splitter *236*
    Output power combiner *240*

8 Standing-wave ratio and related controversies *241*

The electrical length of transmission lines *249*
    The velocity factor of transmission lines *251*
Indicating the effect of reactance in RF systems *252*
"Lost" power in a simple feedline system *252*
Optimum power transfer over a band of frequencies *256*
The destination of reflected power *257*
    Indirect VSWR estimation at the feedline load end *259*
A simple technique to determine line loss *261*
Direct measurement of VSWR *262*
Pitfalls in transferring RF power *264*
    Practical measurement of coaxial cable velocity factor *277*
    The R-X noise bridge *278*

Index *287*

Acknowledgments

THE FOLLOWING LIST OF COMPANIES HAVE BEEN PROMINENTLY IN THE forefront of those specializing in the progress of RF technology. I am gratefully indebted to their assistance in the preparation of *Practical RF Power Design Techniques*: Amperex, Bird Electronic Corporation, Communications Transistor Corporation, Continental Electronics, General Electric, International Rectifier Corporation, MFJ Enterprises, Mirage/KLM Communications, Motorola Semiconductor Products, Inc., Palomar Engineers, RCA Solid-State Division, Siliconix, Inc., and TRW Inc.

Introduction

Whether or not it is stated, an author must be mindful of the type of reader that is expected to benefit from his or her book. *Practical RF Power Design Techniques* is no exception; it, indeed, has been written with a well-focused editorial slant. It has long been the author's observation that engineers, radio amateurs, technicians, and many other workers in electrical and electronic technology often flounder when implementing RF designs, or when interpreting the operating characteristics of RF systems. Because much written material on RF practice exists, this might initially appear to be surprising.

However, much of the available material on the subject is more or less redundant and about half of it is excessively mathematical for everyday situations. As might then be expected, the remainder is too simplistic for optimum use. What appears to be needed, is a practical balance between these extremes. *Practical RF Power Design Techniques* purports to fill this need. More specifically, its highest value should be found as a supplement to other works, for the topic is so vast that any attempt at truly comprehensive coverage is bound to miss the mark. I will feel a sense of accomplishment if you find insight into areas of RF practice that are often given nebulous treatment in general technical literature.

I recognize that the use of solid-state devices in the production and processing of radio-frequency energy is one of the noteworthy achievements of modern electronics. The dramatic improvements imparted to transmitters, receivers, space communications, and to hobbyist endeavors share parity with the somewhat more visible developments that involve computers, calculators, TV games, and stereo systems. Radio frequencies at appreciable power levels also assume importance in such diverse areas as radar, microwave cookers, induction heaters, navigation equipment, welding equipment, and in numerous other scientific and industrial processes. Useful guidance will be found in matters common to these and other applications in discussions of measurement techniques, RF impedance matching, the nature and consequences of reflected power, low- and high-power amplifiers, transmission-line behavior, etc.

1
Transistors versus tubes

BECAUSE VACUUM TUBES HAVE BEEN, AND REMAIN, VIABLE DEVICES IN THE processing of RF power, generalized comparisons are made between tube and solid-state technologies. Many hams, engineers, and others involved in RF work are accustomed to tube circuits and tend to evaluate solid-state applications in terms of their tube experience. Even though solid-state RF applications tend to displace many tube implementations, a well-defined trend uses tubes and transistors cooperatively.

What is radio-frequency power?

From a purist's point of view, all alternating currents (regardless of frequency) radiate some energy into space. So, strictly speaking, radio-frequency (RF) power does not need to be constrained by any limits imposed on frequency. Practically, however, radiation is negligible at the 60-hertz (Hz) power-line frequency. It has been found that at 10 kilohertz (kHz), the radiant energy can be detected across the ocean if a sufficiently large transmitting antenna is used. Actually, even lower frequencies would be sufficient for such long-distance signaling. However, 10 kHz is an audible frequency, and lower frequencies could, among other things, introduce interference with the audio-frequency information that might be conveyed on such radiant energy. And, of course, the physical size of efficient antennas discourages the use of such very low frequencies.

All things considered, the RF portion of the electromagnetic spectrum starts at 10 kHz. Although such a low frequency no longer finds much use for communications purposes, 10-kHz RF power generators are useful in the industry for induction heating, case hardening, and in similar processes. The designers and operators of such equipment know they are involved with a unique domain of electrical power because of the shielding precautions required to prevent radiation. Such inadvertent radiation can interfere with communications or other sensitive equipment.

2 Transistors versus tubes

With higher frequencies, the radiation of energy into space is more readily accomplished with practically sized antennas; indeed, at tens and hundreds of megahertz (MHz), appreciable radiation can occur from the antenna effects of resonant circuits and even from connecting wires.

In the microwave region of the spectrum, a frequency of 100,000 MHz might arbitrarily be defined as the high-frequency limit of RF power. At higher frequencies, the radiant energy begins to manifest itself more as infrared heat than as radio waves. All told, then, RF power can involve alternating currents from 10 kHz to 100,000 MHz. This range corresponds to the wavelength range of from 30,000 meters (m) to 0.3 centimeter (cm).

In addition to communications applications, RF power is used for radar, for cooking, for induction and dielectric heating in industry, for circuit-isolation techniques, and for ionizing gaseous devices. A proposed use is as an intermediate energy medium in the conversion of solar energy from space to terrestrial electrical energy (see Fig. 1-1). Unlike 60-Hz power, RF power is not produced by rotating machinery. Rather, it is generated by electron tubes and, to an ever-increasing extent, by solid-

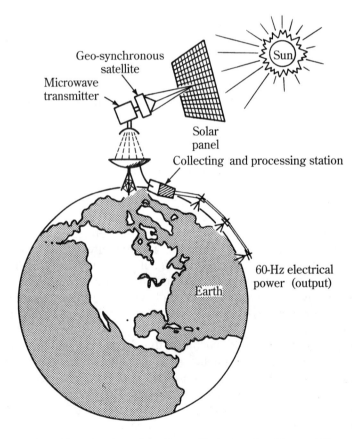

1-1 Possible RF power application of the future. Today's technology would require tubes in the transmitter and solid-state elements at the earth station.

state devices. The relatively recent advent of such solid-state RF power makes a book, such as this, timely.

Why should RF power merit special attention?

When the generation and distribution of electricity first attained commercial proportions, the relative merits of alternating and direct current were a controversial subject. It was soon appreciated, however, that each mode had its own domain of applications and that the behavior of alternating current could be just as precisely predicted as that of direct current. When devices became available to amplify and process voice and sound frequencies, the wonder of it all was taken in stride because these audio frequencies were, after all, the well-understood alternating current, although they encompassed higher frequencies than the 50 or 60 Hz used in power engineering.

At about the same time, another mode of electrical phenomena began to be commercially exploited: radio-frequency power. Although it could be rightfully said that this merely represented more involvement with alternating currents, for practical purposes, it is clear that a whole new branch of electrical technology came into being. Indeed, for many years those who specialized in "power" or in "audio" felt somewhat alienated from the practitioners of the radio art. And the viewpoint was reciprocated!

We might naturally ponder the unique aspects of RF power. These can be tabulated as:

- The radiating characteristic of high frequencies is certainly unique. It is hardly a trivial that energy is no longer confined to the wires and components of an "electric circuit."
- The profound effects of wire length and physical dimensions make necessary a unique expertise for practical implementations.
- The tendency for current flow to concentrate on surfaces requires a modified "Ohm's law" approach to dissipative losses.
- The behavior of substances in strong RF fields is dramatically different from that experienced at lower frequencies.
- The ability to guide, reflect, refract, and concentrate RF energy in beams constitutes an art with no parallel at power frequencies.
- The popular design aid of "breadboarding" of RF power equipment contrasts with the more rigorous approach of the 60-Hz designer, who generally does not enjoy the luxury of experimentation.

So much for the distinctive nature of radio-frequency power. Where is it used? Table 1-1 is a partial list of applications for this mode of electrical energy.

Mechanism of radiation

Because the salient feature of RF power is radiation, do not plunge into the actual circuits for producing such power without some appreciation of the nature of this

4   Transistors versus tubes

Table 1-1. The domain of solid-state RF power.

Aircraft radio	Linear amplifiers
Altimeters	Loran
Amateur radio	Marine radio
Broadband amplifiers	Microwave cooking
Broadcast radio	Microwave communications
Citizen's-band (CB) radio	Mobile radio
Collision avoidance systems	Navigation
Dielectric heating	Plasma generators
Distance measuring equipment	Radar
Frequency-modulated (FM) radio	Radio-sonde service
Garage door openers	Satellite communications
Guidance systems	Television transmitters
Induction cooking	Transponders
Induction heating	Welders
Landing systems	

interesting phenomenon. Although this is a fertile field for exercises in higher mathematics, complex physics, and esoteric philosophic considerations, a simple qualitative explanation is sufficient for most purposes in practical electronics. Figure 1-2A shows the electric field around a dipole antenna. Do not be concerned with the length of the antenna, but only that it is driven by an RF generator or transmitter. The changing pattern of the electric field during the cyclic excursion of the RF current is caused by the mutual repulsion between lines of electric force. Thus, they "push" away from the dipole antenna (Fig. 1-2B). This, in itself, does not lead to detachment from the antenna, for these lines could, conceivably, collapse in reverse fashion from the build-up process. Indeed, this does occur immediately around the antenna in the region ascribed to the induction field. However, if the frequency is high enough, those lines,

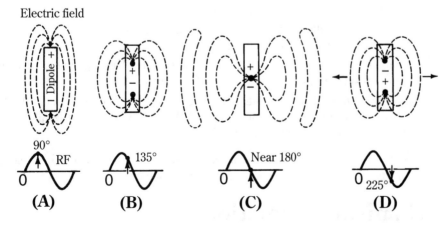

1-2   Production of radiant energy. Electric lines of force which experience insufficient time to collapse before reversal of the current cycle close upon themselves and become detached from the antenna.

which have been pushed sufficiently away from the antenna, are not given enough time to collapse before the charges at the ends of the antenna reverse polarity.

Such electric force lines then do the next best thing; they establish closed paths by closing upon themselves, as shown in Fig. 1-2C. But once having done so, they become detached from the antenna and continue to propagate into space as an electromagnetic wave at the speed of light. The magnetic component is not shown in Fig. 1-2 because the behavior of the electric lines of force is sufficient for explanation of the vital detachment process. The magnetic lines of force are always at right angles to the electric force lines, and therefore can be visualized as always being perpendicular to the plane of the page.

A three-dimensional view of a propagating electromagnetic wave is shown in Fig. 1-3. This illustration depicts the magnitudes of the electric and magnetic fields once the radiant energy has detached itself from the antenna via the previously described mechanism. As shown, propagation is to the right, along the X axis. The speed of propagation in the near vacuum of space is approximately 3×10^8 meters per second (186,000 miles per second). It is slightly less in the atmosphere and can be appreciably less in various dielectric substances, such as water or rock. Although refractive effects of the ionosphere can impart even higher speeds to radio waves, present theory restricts any information or energy carried by radio waves to the previously cited speed; that is, the speed of light in free space.

An important aspect of the propagation scenario is that the electric and magnetic fields wax and wane together, even though the force lines representing them are always mutually perpendicular. If you look closer into this interdependence of electric and magnetic fields, you can conclude that one type of force field begets the other, and vice versa. More specifically, a time-varying electric field produces a time-varying magnetic field. In turn, a time-varying magnetic field produces a time-varying electric field. Because time is required for such fields to go through their respective cycles, a positional advancement of the interaction continually occurs; this constitutes propagation through space.

Why reconsider tubes?

Because this book is intended to impart useful information about RF applications of solid-state devices as power oscillators and power amplifiers, the relevancy of a pre-

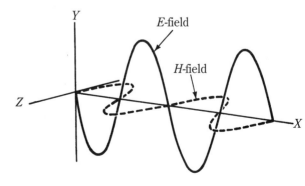

1-3 Spatial representation of an electromagnetic wave. The mutually perpendicular electric and magnetic fields are depicted as propagating from left to right along the X axis.

lude dealing with vacuum tubes might well be open to question. Seemingly, the momentum of solid-state technology has long pointed in the direction of tube displacement. Why, indeed, should it be profitable to rehash tube circuits and techniques?

It so happens that the circuit principles embodied in the new area of solid-state RF applications derive directly from the long-established tube experience. Although it is conceded that it is feasible to "start from scratch" with the solid-state applications, the author feels that a deeper insight into the subject must necessarily stem from acquaintance with or review of tube oscillators and amplifiers. Admittedly, a measure of nostalgia probably permeates this opinion. But for those who are eager to buy the old and become involved with the new, it also happens that tube RF applications are not quite ready yet for burial. In this field, if in none other, the tube retains much vitality; evolving circuit technology often appears in tube designs first, and then is adapted to solid state as devices meeting the contradictory requisites of power capability, frequency, cost, and ruggedness ultimately appear on the market. Tube to solid state represents a continuum in RF power applications, rather than an abrupt transition. Evidence of this is seen in mutual-device systems where, for example, solid-state devices are used in the driver stages and tubes are used in the output stage. Actually, both semiconductor and tube technologies continue to progress. It is clear that the two approaches to RF power can be either competitive or cooperative.

Advantages of tubes for RF work

Despite the manifold features of solid-state devices, tubes possess a few advantages. For one thing, tubes can attain much higher power levels in most RF applications. This remains true even though the power capabilities of today's solid-state devices are at levels that most people would hardly dare fantasize about just a few years ago. Also, tubes are much more "forgiving" than are solid-state devices; they do not readily succumb to catastrophic destruction following abuse. Often the tube comes out favorably in a cost-per-watt comparison (this, however, is no longer construed as an inherent feature of tubes). Until recently, you could say that tube RF amplifiers were easier to drive, but with the advent of the power field-effect transistor, this is no longer necessarily true. Finally, the ability of tubes to operate at high voltages and relatively low currents is, in some respects, advantageous.

Advantages of solid-state RF power devices

It might appear that involvement with this question is tantamount to revival of a topic that was resolved long ago. It, of course, is obvious that solid-state devices have size and weight advantages over their tube counterparts. Of course, dispensing with filament power constitutes an advantage. Perhaps not immediately obvious is the fact that the small size of solid-state devices is advantageous beyond the mere conservation of packaging space. In RF circuitry, particularly at higher frequencies, compact dimensions lead to better circuit performance. Indeed, the physical dimensions of the device, be it tube or solid state, constitute one of the main limiting factors for very high frequency operation. Heat-removal techniques for solid-state devices are more straightforward and are accomplished with simpler hardware. This is because more intimate thermal contact is possible with the heat-developing elements of solid-state devices than with tubes. A big positive feature of solid-state devices is that life-span can be characterized by extreme longevity. Moreover, they tend to die at full performance; usually, demise is not preceded by mediocre performance (sudden death is often the price of abuse!).

Because of physical compactness, solid-state RF devices paved the way for the development of transceivers and truly portable transmitters. This development is also facilitated by the voltage compatibility of many solid-state RF power devices with signal-level solid-state circuitry. The physical ruggedness of solid-state elements also enters the picture here and is particularly advantageous in airborne, space, and military applications. It is also a fortuitous coincidence that solid-state RF power devices can often operate optimally at the dc voltage levels provided by the batteries in automobiles and boats. The relative freedom from maintenance is one of the most compelling attributes of solid-state RF power, and this translates into economic advantage when long-term operation is taken into account.

Graphs such as that in Fig. 1-4 are often cited to indicate the general power and frequency domains of tubes and solid-state devices. Refrain from making a too literal interpretation of such curves, for the performance frontiers of all devices are in a state of flux. At any given time, a few superior performers are not practically available because they have not been mass produced or are reserved for high-paying customers, such as for space programs and the military. Pending an unexpected technological breakthrough, these curves do, however, convey a general picture of the performances that pertain to the indicated devices at a given time. Also, the plotted power levels are continuous, not pulsed.

Class-C tube RF amplifier

If you understand the basic operation of the class-C tube amplifier, most solid-state RF circuits will involve no mysteries. Analogies in performance will, indeed, be closer than you might dare expect. For example, the concept of dualism is often expounded to facilitate comparison between the tube and transistor. According to this idea, the

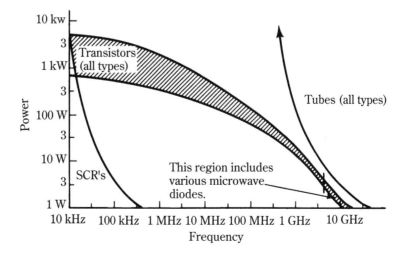

1-4 Approximate power-frequency capabilities of RF power devices.

8 Transistors versus tubes

two devices are similar, but the tube is best viewed as a voltage-actuated device, and the bipolar transistor is most easily explained by considering it to be current actuated. However, with class-C RF amplifiers, the comparison is considerably simplified, because both devices behave as if they were actuated by the input current.

Yet another similarity has to do with the inherent operation of the class-C amplifier. Specifically, it is not an amplifier in the sense that it delivers a power-boosted, but faithful, replica of its input signal. The circuit actually performs as a switch and is used to shock-excite oscillation in an LC tank. Because of this, it really makes no great difference whether the switch is a tube, a bipolar transistor, or a field-effect transistor. Indeed, even an SCR can operate in a similar manner to develop low-frequency RF power in a resonant circuit.

It should now be evident why some time devoted to tube RF circuits is more likely to be rewarding than wasteful. Yet another reason for such an editorial format is the practical fact that solid-state RF circuitry is often deployed to replace tube designs. It will surely enhance your expertise to know the nature of the displaced circuit and device.

Figure 1-5 depicts a representative class-C RF amplifier using a triode tube. Similar output stages are used in numerous transmitters. Providing that the required power output level is not too great, an ever-increasing number of such transmitters are destined to be manufactured with solid-state output stages. In any event, a salient feature of the circuit of Fig. 1-5 is the application of sufficient negative grid bias to more than cut off plate current flow. This being the case, a large input signal must be impressed on the grid to overcome the inhibiting effect of the high negative bias. In actual operation, only the positive tips of the input signal turn the tube on. The plate current flows only in short-duration bursts. Thus, the tube is caused to simulate a switch, which, although synchronized to the input signal, is

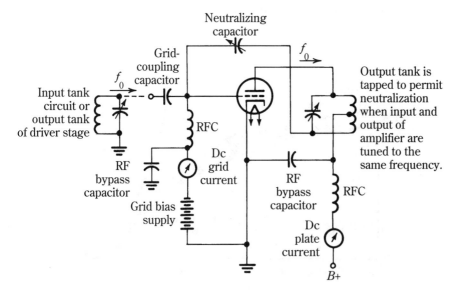

1-5 Representative tube class-C RF power amplifier. Reconsidering the operating principles of this basic circuit provides insight into the newer solid-state systems.

allowed to be on for only a small fraction of a complete cycle. The term "amplifier" is clearly a misnomer.

Basic operating features of class-C amplifiers When hams and RF circuit engineers made the transition from tubes to solid-state devices, they surprisingly discovered that they were right at home with the basic operating features of class-C amplifiers. For example, very high efficiency still prevailed, and the counterpart of the plate meter still obligingly dipped to indicate resonance during tune-up. These two operating features were not always well understood during the long domain of the tube. In retrospect, it is easy to perceive that neither feature stemmed directly from the nature of the tube itself, but rather from the overall class-C circuitry.

The waveform diagram of Fig. 1-6 provides useful insights into class-C operation. Plate current has a low duty cycle; that is, plate current is off for longer intervals than it is on. Although this suggests the possibility of high-efficiency operation, a little study of waveforms reveals an even more important operating characteristic, which must lead to low power dissipation, and therefore high efficiency. Observe that, with the output LC circuit resonant at the input frequency, the plate voltage is minimum during the very time that plate current is maximum. Because the product of these two

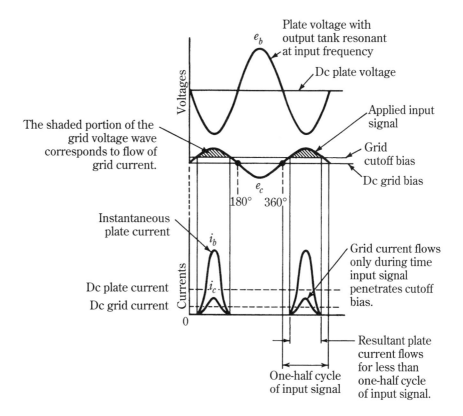

1-6 Voltage and current relationships in the tube class-C power amplifier. Obviously, the term "amplifier" is a misnomer, because the plate current is not a replica of the input signal, but is, rather, a sequence of short-duration pulses.

quantities is plate dissipation power, this all-important tube loss is inherently low. Once this basic operating mode of class-C tube amplifiers is grasped, it can readily be appreciated that the circuit should provide essentially similar operation, regardless of the physical nature of the switch. In other words, it should not make a great difference whether the dc pulses that shock-excite the LC tank are delivered by a vacuum tube, a bipolar transistor, a field-effect transistor, or another device with a control electrode.

The plate-current dip at resonance is one of the widely known behavior characteristics of class-C amplifiers. Yet there is considerable misunderstanding with regard to its cause. Here again, the waveforms in Fig. 1-6 are informative. It is evident that the dc plate-current meter must indicate some kind of average value because the plate current is actually in the form of narrow pulses. From the Fourier theorem of wave composition, such a pulse train consists of a dc component (zero frequency) plus numerous harmonics. The first few harmonics have considerable effect on the shape of the pulses and, therefore, their average dc value. In other words, the dc component, itself, is the result of the magnitude and phase relationships of the harmonics. Moreover, the first harmonic, which is the *fundamental frequency*, exerts the greatest effect on the pulse waveform. When the LC tank is resonated, the maximum impedance to the flow of the fundamental-frequency currents prevails. This alters the shape and magnitude of the plate-current pulses in such a way that the net dc component (the average value of plate current actuating the plate-current meter) is greatly reduced. Thus, the situation becomes somewhat of a paradox in that attenuation of one of the Fourier components (the fundamental frequency) reduces the net dc component of the plate-current waveform. Accordingly, the plate-current meter obligingly dips.

Linear RF amplifier

Not long after tube RF amplifiers began to be deployed for achieving high power levels, it occurred to an enterprising amateur that his amplitude-modulated transmitter could develop more power by the simple expedient of inserting a large class-C amplifier between the present modulated stage and the antenna. Such a final amplifier did boost the RF power, but it practically destroyed the fidelity of the modulation. With the combination of empirical and analytical techniques, a way was found to produce the sought-after boost in RF power level without distorting the modulating signal. After this was implemented, the circuit of the class-C amplifier remained virtually unaltered. What was done to bring about the desired operational mode?

The class-C RF amplifier cannot properly handle a modulated signal impressed at its input because the current produced in its output circuit is not proportional to the voltage applied to its input circuit. This could probably be anticipated from the fact that such an "amplifier" is, in essence, more in the nature of a pulsing or switching system that completely depends on the energy storage in its output tank circuit for the production of undamped, constant-amplitude, RF waves. It turns out that some aspects of the class-C amplifier can be retained and still have the requisite proportional amplification needed to properly boost the power of modulated waves. Specifically, if the input excitation consists of half cycles of the RF wave, this important operating condition will occur. That is, instead of supplying 90° to 120° (for example) of the RF sine wave to the amplifier's input, 180° (one-half cycle) is impressed. To accommodate this change, the negative grid bias must be changed; it must be lowered in order to enable the tube to operate as a proportionate, linear, amplifier over the entire half-cycle of impressed RF. Thus, the circuit itself can conceivably remain unchanged. This is reminiscent of class-B audio amplifiers, but the comparison is faulted by the strange fact that one tube is sufficient to produce essentially undistorted output of the modulated RF wave. How can this be so, when it is obvious that severe distortion would accrue in a class-B audio amplifier if one of its push-pull power tubes was removed from its socket?

The resolution of this paradox lies in the tuned LC tank circuit of the class-B RF amplifier. Its energy storage, its tendency to "ring," supplies the missing half of the RF cycle. This enables the tube to impart linear amplification of the modulated wave. The class-B RF amplifier thus retains the salient feature of the class-C amplifier in that a tuned circuit supplies output power when the input drive no longer is present. At the same time, the plate current is zero with no input so that the low operating efficiency of class-A operation is circumvented. The theoretical peak efficiency of a class-B linear amplifier is 78.5%; that of a class A stage is 50%.

More-practical linear amplifiers

In practice, it is desirable to operate the linear RF amplifier in the class-AB region, rather than "purely" in class B. This improves the linearity and generally holds true for solid-state, as well as tube, stages. The linearity improvement stems from two sources. First, the dynamic transfer characteristic of the amplifying device is "straightened out." Second, a more constant load is presented to the driver stage. Of course, the "blend", with class-A characteristics degrades the operating efficiency somewhat. In actual practice, it is feasible to achieve a good compromise. This is often brought about by adjusting the bias so that some plate current exists without any RF drive. Such an idling current is, however, much smaller than would exist if the stage was biased for class-A operation. The overall result is linearity approaching that of a class-A amplifier and operating efficiency not drastically reduced from that attainable with class-B operation.

Even better results are generally forthcoming when the tube is operated in the grounded-grid configuration. Here, the control grid is at ground potential (for RF), and the drive is applied to the filament or cathode. At low and medium frequencies, up to 20 or 30 MHz (for example), the electrostatic shielding action of the grounded grid often stabilizes amplifier operation so that neutralization can be dispensed with. Even when neutralization is used, less critical adjustment and extended broadband operation generally result. Although the impedance presented to the driver stage is low, it is more constant than with grid-driven circuits. The low input impedance of the grounded-grid stage actually makes it more convenient to use coaxial cable between the driving stage and the final amplifier. A modem high-power grounded-grid amplifier, such as shown in Fig. 1-7, can be excited from a solid-state driver.

Although the input to the grounded-grid amplifier is a power-consuming circuit, most of the power thus consumed from the driver stage is not dissipated as loss. Rather, it reappears in the output circuit and actually adds to the output power. Typically, 10% of the output power actually derives from feedthrough from the driver stage. This is not of consequence in CW, FM, or SSB operation, but it will prevent

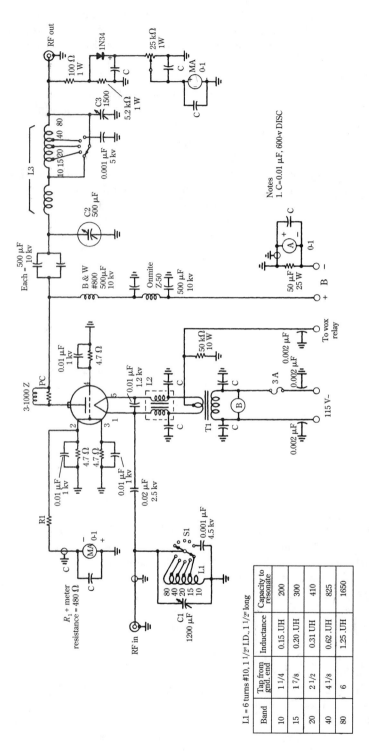

1-7 Typical circuit of high-power grounded-grid amplifier. Numerous radio amateurs, as well as designers of industrial RF equipment, have derived much of their experience from circuits such as this.

100% modulation of an AM signal. The grounded-grid circuit is usually a more effective amplifier for the VHF and UHF regions than the "conventional" grounded cathode configuration. Familiarity with tube grounded-grid amplifiers pays dividends when working with grounded-base transistor circuits; many operating characteristics are quite similar for the two devices in these circuit configurations.

Transistor version of the grounded-grid amplifier

Seemingly, a transistorized counterpart of the grounded-grid amplifier would be a "natural," if for no other reason than that neither filament transformer nor filament chokes would be required. You might also be enthusiastic over the better frequency response inherent in common-base circuits, as compared to the common-emitter configuration. Also, you might have been accustomed to seeing superior linearity in power transistors characterized for audio-frequency service when such transistors are operated in common-base amplifiers. Except for certain servo applications, common-base power amplifiers are now rarely encountered. Although common-base RF amplifiers (and oscillators) are used, especially at VHF, UHF, and microwave frequencies, most transistor power amplifiers, whether class A, B, C, or "linear," are configured about the common-emitter circuit.

The main reason that transistor versions of the grounded-grid amplifier are not commonly found is that the input impedance of such an arrangement would be so low that practical difficulties would attend the design of an input matching network. Also, the linearity of RF power transistors generally offers little, if any, advantage when operated in the common-base circuit. Whereas a tube tends to have better RF stability in grounded-grid than quite grounded-cathode circuits, the situation prevailing with transistors might be different; that is, RF stability is often readily attained in common-emitter, but not necessarily in common-base, amplifiers. Notice that although the grounded-grid amplifier generally dispenses with neutralization, the common-emitter amplifier generally requires no neutralization. Inspection of many common-emitter amplifiers shows that neutralization is usually not required. This is because of low Q "tank" circuits, deliberately introduced emitter degeneration, and the low gain of transistors at high frequencies.

There are, however, similarities between tube and transistor linear amplifiers. Both are operated as close to class B as is consistent with acceptable linearity. In practice, this usually means that both tube and transistor "linears" operate in the class-AB region. That is, both consume small dc idling currents when there is no RF input.

What has been said about transistor versions of the grounded-grid amplifier applies primarily to the bipolar transistor. The relatively new power FET might find service in the analogous circuit (common gate) to the grounded-grid configuration. However, thus far the linearity displayed in common-gate arrangements does not provide much incentive to simulate tube practice.

Solid-state approach to high power

A variety of RF power transistors are available that can develop power outputs of 50, 75, and 100 watts (W) and even greater. Many of these transistors can be readily accommodated in broadband circuit arrangements so that the 1.8- to 30-MHz frequency range can be handled without tuning adjustments. The power supply required generally ranges from 12 to 28 volts (V). This is respectable power for many purposes, and somewhat more power is available via push-pull and parallel implementations. But, suppose you want power at the 1-kilowatt (kW) level. The tube-oriented amateur or designer would probably consult a semiconductor firm's catalog in quest of an appropriate giant transistor. The odds are against finding one. Although kilowatt transistors have been made for military and industrial RF applications, it is just as well that they are not a common commodity on the market.

The practical problems attending the use of such a giant transistor are far from trivial. Conductors, coils, and RF chokes capable of carrying the tremendous dc current would be so massive as to pose serious construction problems. It is not merely a matter of attaining safe current-carrying capacity, but rather of avoiding power-deleting voltage drops. At current levels of, 150 (for example) amperes (A), "small" resistances on the order of several tens of mΩ would seriously impair the operating efficiency. Aside from this annoying situation, it might not be such a good idea to depend on a single device. Thinking of the ease with which lower-power RF transistors can often be destroyed during experimentation or tune-up, it could be less than pleasant to blow out one of the giant units.

It appears wiser to try to achieve high power by somehow combining an appropriate number of smaller transistors. Paralleling is generally not desirable with bipolar transistors. Troubles with current hogging quickly occur; besides, such a

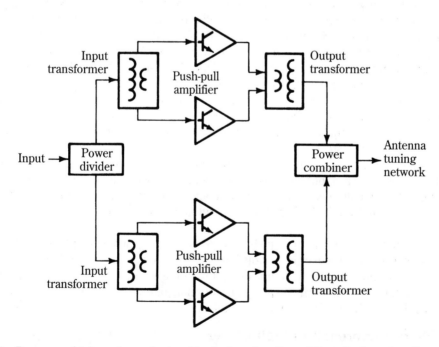

1-8 Power-combining scheme for two identical push-pull amplifiers. Each push-pull amplifier is complete in itself and operates independently of the other. The antenna load receives the summed power of these amplifiers or four times the output power of a single transistor. The power divider and combiner are special transformers.

combining method does not represent good RF practice at the higher frequencies. This is just as well; if it was feasible to parallel a large number of 80- or 100-W transistors to develop a kilowatt output level, you would be back at the beginning with contemplating the single giant transistor. Push-pull circuits are effective for at least doubling power, and in some, but not all, cases a push-pull and parallel arrangement of four transistors can be successfully worked out. Such a scheme would probably yield about 3.5 times, rather than 4 times the available power from one transistor. There must be a better way!

Fortunately, there is another approach. Power-combining circuits can be implemented in which two, three, or four pairs of push-pull transistors have their individual power outputs summed in the load. The block diagram of Fig. 1-8 depicts such a technique for two separate push-pull amplifiers. The load power is four times the contribution of any of the individual transistors. The *power divider* (also called a *splitter*) and the *power combiner* are similar to one another and are special *hybrid transformers*. They have their windings arranged and connected so that the amplifier inputs are isolated from one another in the case of the divider, and the amplifier outputs are isolated from one another in the case of the combiner. These transformers are commercially available.

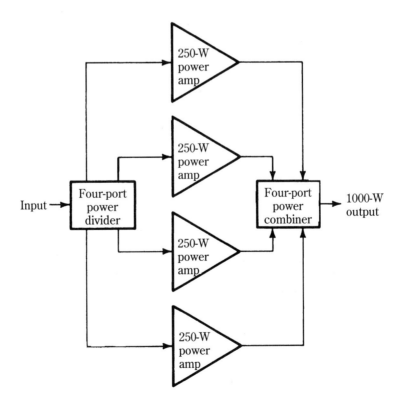

1-9 Extended use of combining technique for producing still greater load power. Here, the power-dividing and power-combining transformers each have four ports. Moreover, each 250-W RF power amplifier can, conceivably, be a complete subsystem, such as shown in Fig. 1-8.

16 Transistors versus tubes

The power-combining concept depicted in Fig. 1-8 can be extended to produce even greater power from commercially available RF power transistors. For example, divider and combiner hybrid transformers can be designed to have four ports (Fig. 1-9). This enables the summing of power from four separate RF power amplifiers. Interestingly, each RF power amplifier shown in Fig. 1-9 can be almost any amplifier that can stand alone and perform well. For example, each of these amplifiers can represent an entire subsystem, such as the entire arrangement of Fig. 1-8. A practical and economical feature of this technique is that the eight push-pull amplifiers that would thereby be deployed in the technique of Fig. 1-9 would be of identical circuitry and construction. Yet another feature is that the loads of either Fig. 1-8 or 1-9 continue to be supplied with proportionate power if one or more of the individual amplifiers fail. Such operational redundancy would certainly be conspicuous by its absence if you used a single giant transistor.

Wilkinson power-combining technique

The *Wilkinson power-combining technique* provides a simple and straightforward method to combine the power capabilities of two amplifiers where broadbanded operation is not needed (Fig. 1-10). The elements used are four quarter-wave transmission lines. Although initial inspection of the arrangement might suggest push-pull operation, the amplifiers are essentially paralleled. But, unlike ordinary paralleling, there is considerable electrical isolation between the two amplifiers.

Assuming a 50-Ω input source, a 50-Ω load, as well as 50-Ω input and output impedances of the amplifiers, the quarter-wave lines should have a characteristic impedance, Z_0, of $50 \times \sqrt{2}$, which is approximately 70 Ω. This being the case, you can show that matched impedance conditions prevail. For example, the line A associated

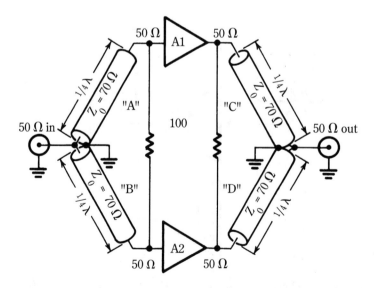

1-10 Wilkinson power-combining technique. Under ideal conditions, the 100-Ω resistances dissipate no power.

with the input of amplifier A1 steps its input impedance of 50 Ω up to 100 Ω at the input terminal of the pair of amplifiers. This comes about from the impedance relationships of the quarter-wave transformer, in which $Z_0 = \sqrt{Z_1 \times Z_2}$. Substituting the relevant values, $70 = \sqrt{100 \times 50}$, Z_1 and Z_2 being the terminating impedances of the line. Because two such lines join at the input of the amplifier pair, the actual input impedance seen by the driver or source is 100/2 = 50 Ω. The same basic reasoning applies to the output circuit involving the amplifier outputs, lines C and D, and the load. Notice, however, that the input circuit has power splitting or dividing, whereas in the output circuit, power combining occurs.

When the impedance situation is as described, no power dissipation occurs in the 100-Ω resistances. If ideal impedance conditions do not exist, these resistances will absorb the reflected power that will thereby be produced.

More than two amplifiers The hybrid networks involving two quarter-wave lines for summing the powers of a pair of amplifiers are actually a specific application of a more generalized technique. Any number of amplifiers can be handled in this manner; it is merely necessary to extend the scheme to accommodate the amplifiers involved. Figure 1-11 shows the more generalized arrangement for use with n, that is, any number of amplifiers. Again, the network can be deployed either as a power divider or power combiner. Notice that the configuration reduces to that of Fig. 1-10 when $n = 2$. R_0 represents the source (driver) and load impedances and is

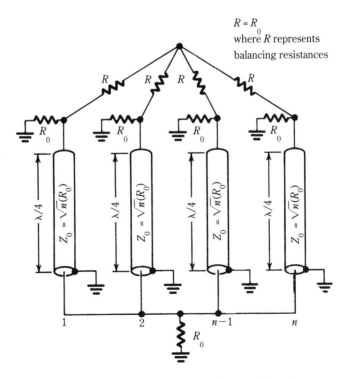

1-11 Wilkinson divider-combiner network for use with n amplifiers. R_0 represents the source and load impedances, generally 50 Ω.

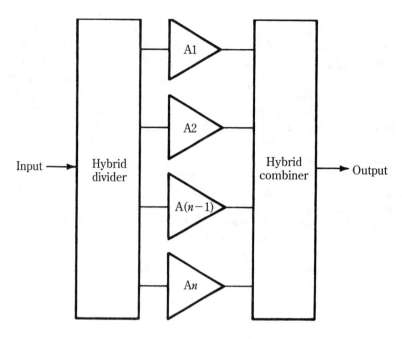

1-12 Wilkinson hybrid system with n amplifiers. A general power-combining situation is depicted by the block diagram.

usually 50 Ω. The characteristic impedance of the lines, Z_0, changes with the number of lines used. Other parameters and considerations remain the same, regardless of the number of lines. Generally, a 25% bandwidth is attained, and the isolation between ports is on the order of 25 decibels (dB).

The block diagram of Fig. 1-12 shows the generalized method of summing the outputs of any number of amplifiers with the use of power dividers and power combiners that have the basic configuration depicted in Fig. 1-11. Interestingly, the amplifiers themselves can be of any format. That is, they can be configured around single transistors, point-to-point paralleled transistors, push-pull transistors, or power-combining systems so that hybrid systems are within hybrid systems. Generally, all amplifiers from A1 through An are identical, but it is conceivable that amplifiers of diverse formats can have their powers summed in the load via this scheme.

As described, this power-summing method is suitable for narrowband applications. However, considerable broadbanding can be accomplished by using stepped stripline quarter-wave elements, or by cascading techniques, in which the lines become odd numbers of quarter-wave lengths beyond 1. Both the application and the mathematical relationships might become unwieldy, however.

Amplifier module

Amplifier modules are relatively new, but are already a widely used solid-state RF product. Although various fabrication techniques are used, the common feature of these power amplifiers is that the installation consists merely of inserting the unit between the driver (usually the previously used final amplifier) and the antenna, and

Why reconsider tubes? 19

1-13 A typical RF power amplifier module. Not visible are input and output BNC connectors for 50-Ω coaxial cable.

providing a dc source of power (usually a nominal 12 V). They are sufficiently broadbanded for their intended purpose so that no tuning adjustments need to be made. Indeed, there is generally no external adjustment or tuning provision, other than a toggle switch or two for turning on and off the dc power and selecting the operational mode (class B or AB for single sideband, class C for CW or FM). These modules have thus far found greatest application in the VHF and UHF bands where stripline "tank" circuits enable predictable performance from mass-produced units and contribute to stable operation. A large market for these amplifier modules have been the owners and operators of transceivers who are seeking a convenient and inexpensive way to boost power. Power outputs from 15 to 150 W have been commonly available.

The fabrication techniques use various blends of hybrid and discrete methods, with the heatsink often being the most evident construction feature (see Fig. 1-13). Manufacturers vie with one another to provide various performance features. Among these are output harmonic filters, integral preamplifiers for easy drive, transmit-receive relays, jacks for monitoring performance, and, above all, high efficiency. These amplifier modules are highly useful for mobile operation and are similarly being welcomed for use in marine and aircraft applications.

Some amplifier modules are intended for use in MATV/CATV installations. These generally have power-output levels of a fraction of a watt to several watts and might exhibit substantially flat response from 40 to 300 MHz. With somewhat lower power ratings, there are other amplifier modules that are broadbanded from 1 to 250 MHz or from 10 to 400 MHz. It is remarkable that, despite the sophisticated technology and careful design incorporated in RF amplifier modules, they can almost be treated as just another component insofar as concerns installation and operation.

Low-power RF amplifier module

Hybrid-constructed amplifier modules are useful for driver, predriver, and signal-processing functions in more extensive systems. Their deployment saves engineering development time, conserves space, and often provides optimized performance features that are not readily attained with discrete circuit layouts. Their consideration is worthwhile in light of the fact that most circuit schematics of power-output amplifiers merely depict an IN terminal, which thus evades the problem of excitation.

The physical appearance of such a module, the Motorola MHW591, is shown in Fig. 1-14. The maximum dimensions are such that the module occupies a volume of less than 1.2 cubic inches. It nonetheless contains eight transistors and three transformers. An alumina substrate is used with thin-film techniques, which include nichrome resistors, gold strip-line elements and interconnections, molybdenum heat-spreaders (for thermal conductivity), and chip capacitors.

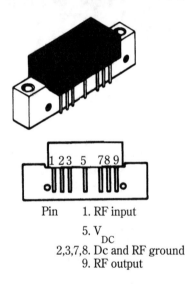

Pin   1. RF input
5. V_{DC}
2,3,7,8. Dc and RF ground
9. RF output

1-14 Example of hybrid-constructed RF amplifier module. Intended for driver, and other low-power applications, such modules take up little space, involve high reliability, and often have certain optimized performance features.
Motorola Semiconductor Products, Inc.

The output power is on the order of 1 W and linearity is exceptionally good. Its overall characteristics prevail when the module is used with 50- to 100-Ω impedance levels. Probably the salient feature of this particular amplifier module is its inordinately broadbanded response (Fig. 1-15). Such a module is, itself, a well-engineered subsystem, and it is difficult to identify its counterpart in RF designs that use tubes. As might be surmised, the uniformity, stability, and reliability of hybrid modules of this type are especially noteworthy.

Reliability   The reliability of a transmitting-type vacuum tube is, of course, subject to many variables and vagrancies. However, we are reconciled to the fact that its life span is primarily governed by the electron-emission ability of its filament or cathode. In other words, much as with an automobile tire, the device is eventually "used up." As with the tire, its functional usefulness generally diminishes from the moment it is pressed into service. Those making the initial transition to solid-state RF power often tend to assume that concern with life span is an archaic notion, and maintenance-free operation is the natural mode of performance with modern silicon devices. Such

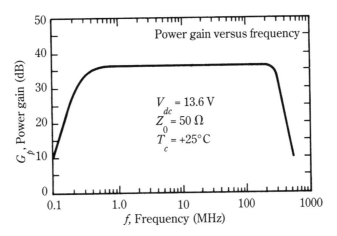

1-15 Broadband frequency response of the Motorola MHW591 hybrid amplifier model. Such optimized performance features are not readily attained via the more traditional construction techniques. <small>Motorola Semiconductor Products, Inc)</small>

an assumption might, indeed, appear justified in many practical situations. On the other hand, disappointment might be in store for the subscriber to the premise of indefinitely great longevity for solid-state RF power devices. Besides being vulnerable to catastrophic destruction from transients, "hot-spotting," overdriving, and the like, semiconductor components actually have built-in wear-out mechanisms.

One wear-out mechanism is *metal migration*. With high current density and high temperatures, the metallic ions in the interconnect or wire bonding system actually are displaced. This is somewhat reminiscent of electroplating, where ions in a solution are deliberately imparted motion by an electric current. This phenomenon is relatively slow in metals, but eventually produces whiskers, hillocks, and voids. Failure in the device then manifests itself as an internal short or open circuit. Gold is better than aluminum metallization in delaying the onset of this failure mode, and monometallization is generally superior to fabrication processes that use more than one metal. This, however, is influenced by the nature of the metals that might be in contact with one another.

A closely related failure results from corrosion of metallic bonds, contacts, and interconnects. The semiconductor material, itself, might contribute to such failure.

Another well-documented failure mode in all power semiconductors is *thermal fatigue* of leads and interconnect metallization. This represents accumulated mechanical strain from the kind of temperature cycling, which occurs when the device is first turned on or finally turned off.

Certain abusive conditions during operation or adjustment can shorten the natural life span without necessarily incurring any of the preceding failure modes. Included are overdrive (particularly where the emitter-base diode is driven into avalanche or zener breakdown), high VSWR in the load, voltage transients from the dc supply, parasitic oscillations, and high junction temperatures.

Figure 1-16 shows failure probabilities for a family of RF power transistors as a function of junction temperature. The sample calculation depicts a mean time to

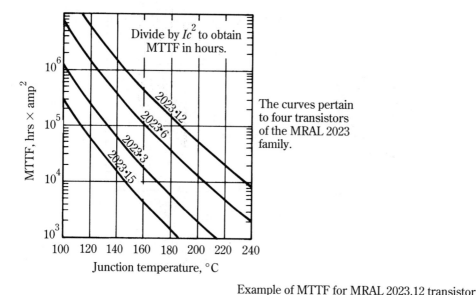

1-16 Typical RF power-transistor failure prediction chart. The accompanying sample calculation shows that good, but not infinite, longevity may be expected. *TRW Semiconductor*

failure of about 26 years for typical operating conditions for one of the types. This is a long expected life span for hobbyist and consumer applications, but not necessarily for space or military uses. And you need not be particularly skilled at extrapolation to interpret lifespans of 5 or 10 years for other operating conditions. Considering that the chart predictions are merely weighted averages and that the transistor is always subject to random deviations that are appreciably removed from the "mean" probability value, it is obvious that the subject of solid-state reliability deserves consideration. This is especially true because solid-state devices usually suffer catastrophic destruction when they fail. From a maintenance point of view, you cannot rely on any internal time clock, as with thermionic emission in a vacuum tube.

Although only the highlights of a very extensive subject has been touched upon here, remember one overwhelmingly important factor in the interest of life span: junction temperature. Although collector-base junction temperature is implied, it is not to be inferred that high operating temperature is of benign consequence in the emitter-base junction. However, in most applications, the power dissipation in the emitter-base junction is much less than in the collector-base junction. Therefore, the temperature of the emitter-base junction tends to follow that of the collector-base junction. When hot-spotting, or base emitter avalanche, occurs, instantaneous or small-area temperatures in the emitter-base junction might assume destructive levels without appreciably affecting the average junction temperature.

Summarizing, the insurance agent might have a different conception of semiconductor longevity than the complacent circuit designer. Transistors are not as forgiving as are tubes; it is beneficial to devote as much time to heat removal as to circuitry to spare them from power-supply transients, to never apply excessive drive power, and to try to provide a low VSWR matched load at all times. A final suggestion, one that all too often is derived from painful experience, consists of procuring a device in which the manufacturer takes pride in quality control and reliability screening.

Test circuits

Most manufacturers of solid-state RF power devices provide test circuits in conjunction with parameter values, tables, and graphs. Such circuits depict typical situations for which the technical information is valid. Both maker and user benefit from these circuits. The maker is largely spared from allegations of unrealistic performance, and the user is provided at least initial guidance, which is predicated on good engineering practice.

A typical test circuit is shown in Fig. 1-17. Here, the manufacturer illustrates a simple, yet sophisticated, application for a unique product, the JO 2058 transistor, which is specially designed to operate in the commonbase configuration and to produce clean 100-W pulses of 450-MHz energy. Anyone with a modicum of experience in high-frequency and pulse techniques and who has worked with RF at high power levels and low impedances can appreciate the likelihood of going astray without such basic "navigation." Indeed, many designers find that it is good practice to spend a little extra time ferreting out a manufacturer's test circuit, which closely complies with the requirements of the application at hand. Minor modifications will then be easier and safer to incorporate than will major ones. This is much more true with transistors than with tubes. Most RF power transistors are optimized for best performance under specific circuitry and operating conditions; forcing them to perform under other circumstances can, at best, result in diminished life span and, at worst, produce instant catastrophic destruction.

Layout diagrams and PC-board art

At the higher frequencies, a connection diagram is not sufficient information to enable reproduction of the test circuit. This is especially so at VHF, UHF, and microwave

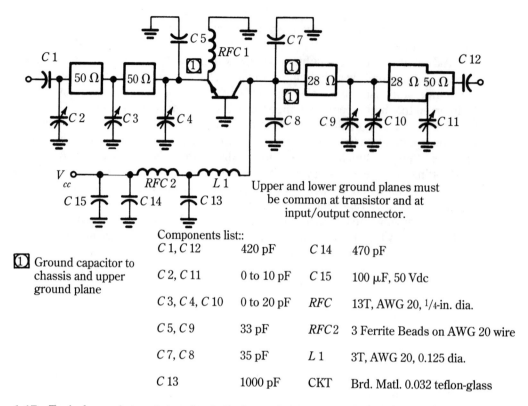

1-17 Typical manufacturer's test circuit. Such test circuits often provide the basic design information needed to implement a device. This circuit is that of a 100-W pulse amplifier intended for production of 1-ms pulses of 450-MHz energy. The transistor is the JO 2058. _{TRW Semiconductor}

frequencies where stripline techniques are used as a format for the elements in the impedance-matching networks. Therefore, it is also commonplace for the manufacturer to back up test circuits with layout drawings and with printed circuit (PC) board artwork. Figure 1-18 illustrates such pictorial data for the test circuit of Fig. 1-17. The importance of this procedure stems from the fact that the impedance, reactance, and Q of the stripline elements are governed by their physical dimensions and by the nature and thickness of the PC board. Thus, the oft-repeated admonition to keep leads short in RF circuitry is not, in itself, sufficient for reproducibility of performance where the "leads" are deliberately designed to yield circuit functions.

Amplifier chains

In solid-state RF power, a practice has come into prominence, which although not unknown in vacuum-tube techniques, was not commonplace. This pertains to device manufacturers's concern with the system, as well as the individual amplifier stage. To this end, "amplifier chains," or "suggested lineups," are often included in the specifications literature. Figure 1-19 is a typical example of such information. From the user's viewpoint, this can save considerable intellectual and empirical effort. From

Why reconsider tubes? 25

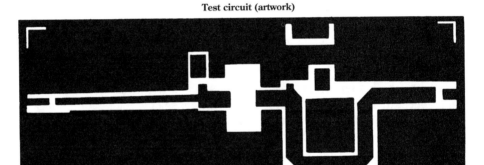

1-18 Board layout and artwork for the test circuit of Fig. 1-17. The physical and geometrical aspects of such a circuit are as important as the schematic diagram. <small>TRW Semiconductor</small>

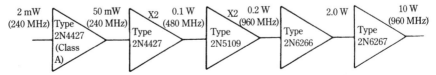

1-19 Typical amplifier chain. Such suggested stage lineups provide the designer and experimenter with relevant guidance at a glance. <small>RCA Solid-State Division</small>

Low-frequency applications

The 10- to 500-kHz region composes the low-frequency portion of the RF spectrum. Rationale for such classification is simply that the preponderance of technical literature addresses itself to techniques pertinent to much higher frequencies—from several megahertz through the far-microwave region. Low frequencies are readily generated and processed by traditional oscillator-amplifier lineups via the use of power transistors. Below 30 kHz, it often is the case that SCRs are more desirable than transistors—especially for high-power final amplifiers. Many designs use class D, rather than class C or linear operation. Here, the output voltage wave is square and the duty cycle is ideally 500%. This leads to very high efficiency, as well as to better reliability. Such operation is feasible because of the excellent performance of harmonic filters in this frequency range. Not only do such filters provide high attenuation of harmonics before they can reach the antenna, but arrangements are readily made whereby the harmonic energy of the square wave can be rectified for use as auxiliary dc power. This, of course, further enhances the overall operating efficiency of the transmitter.

Many are not accustomed to think of frequencies of, say, several tens of kilohertz as radio frequency. However, all the attributes that are ordinarily associated with RF exist: resonant circuits are used, radiation occurs both intentionally and inadvertently, and high-frequency phenomena, such as dielectric and induction heating, manifest themselves. In fact, one of the early uses of wireless communications was to span the Atlantic Ocean with frequencies of 12 to 17 kHz. More modern applications involve very powerful transmitters in about the same frequency region, which provide one-way communications to submerged submarines. The very low frequencies actually comprise a gray area where the line of demarcation between power, audio, supersonic, and radio frequencies is somewhat muddled. For example, a frequency of, say, 19 kHz can be shown to possess power, supersonic, and radio-frequency characteristics, all the while displaying circuit attributes familiar enough to the audio buff.

In any event, low frequencies in the 10- to 500-kHz region are commonly found in the following applications:

- Radio transmitters
- Ultrasonic cleaning and mixing
- Ultrasonic welding
- Sonar transmitters
- Switching power supplies
- Inverters
- Dc-to-dc converters
- Induction-heating generators and cooking ranges
- Fluorescent lighting supplies

Low-frequency applications 27

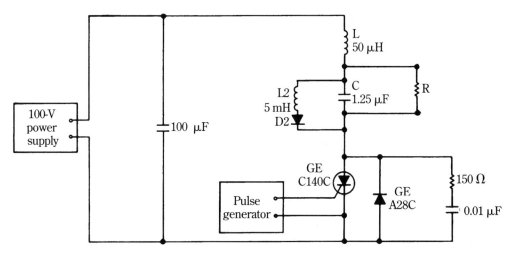

1-20 Basic inverter circuit for producing low-frequency power. With appropriate selection of SCR and components, several kilowatts of low-frequency (to 30 kHz) power can be developed in load). _{General Electric Co.}

Notice that these applications can involve appreciable power levels, from several tens of watts to many kilowatts. The previously mentioned submarine communicator operates at peak powers in the vicinity of 2 MW. Those working with the applications listed must deal with such factors as Q of LC circuits, RFI and EMI, harmonic suppression, shielding, shock excitation, parasitic oscillation, frequency stabilization, neutralization, and modulation formats. Thus, you must deal with radio-frequency techniques—even if the objective is to produce a dc switching power supply.

Basic SCR inverter

The circuit shown in Fig. 1-20 exemplifies a simple arrangement that enjoys widespread use in low-frequency RF applications. The SCR is used to shock-excite the series LC circuit, but the output frequency is governed by the trigger pulse rate that is applied to the SCRs gate. Thus, the output waveform is not necessarily a continuous sine wave. Figure 1-21 illustrates the output waveform for different trigger rates.

In the circuit of Fig. 1-20, L2 serves to help commutate the SCR. Diode D2 breaks up any resonance that might exist between L2 and C. This is most likely to occur at low trigger rates. L2 is usually chosen to have 10 to about 100 times the inductance of L. In some circuits, depending on the trigger rate, and upon the type of SCR, L2, D2, or both of these components might be dispensed with. In any event, the SCR used in this circuit is of the inverter type or other fabrication that is specifically intended for high-frequency performance. In SCR technology, "phase-controlled" types are designed for use in 60- to 400-Hz application. Units intended for operation from several kilohertz to several tens of kilohertz are considered "high-frequency" types and are commonly found in various inverter circuits.

The output waveforms of Fig. 1-21 can be used in the inverter-driven fluorescent-lamp circuit of Fig. 1-22. Here, excellent control of the light intensity is, obtained by

28 Transistors versus tubes

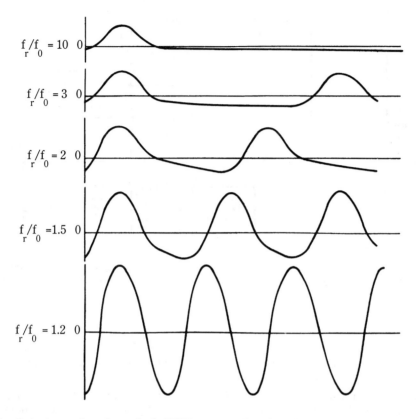

1-21 Output wave form from single-SCR inverter as function of trigger frequency. Sinusoidal wave shape is approached as trigger frequency, f_0, nears LC resonant frequency, f_r.

varying the frequency of the gate triggers. This is because the voltage pulses delivered to the fluorescent lamps remain high enough for ionization—even at very low rates of gate triggers. Because the duty cycle of the pulses delivered to the lamps is low under such conditions, the average intensity of light output will be low. There will, however, be no visible flicker because the repetition rate is still far beyond visible perception. LB designates inductive ballasts. These are needed because of the negative-resistance characteristic of the lamps. They help maintain essentially constant lamp current. Although the nominal operating frequency of this circuit has been in the 5-kHz region, modern SCRs enable frequencies in the 15- to 20-kHz region to be conveniently used.

Three methods of coupling the output of the single-SCR inverter to ultrasonic transducers are shown in Fig. 1-23. Ultrasonics applications are important in cleaning, mixing, welding, and in other industrial processes. Sonar and depth-measuring equipment, likewise, make use of such acoustic radiation. SCRs provide good performance up to about 30 kHz. At higher frequencies, power transistors and power Darlingtons are either competitive or superior. Some ultrasonic transducers operate in the megahertz range, where SCRs are no longer suitable. It is interesting to con-

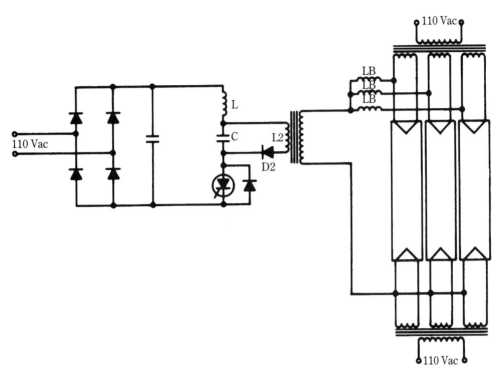

1-22 Inverter-driven fluorescent-lamp circuit. Dimming is accomplished by lowering the frequency of trigger pulses applied to gate of SCR. _{General Electric Co.}

template that even low-frequency ultrasonic waves in air and in fluids have wavelengths comparable to high-frequency electromagnetic radiation. That is why such phenomena as cavitation, reflection, refraction, and focusing into beams are commonly encountered. Thus, even at 15-kHz or so, the designer or experimenter versed in RF techniques finds himself or herself in familiar territory.

Yet another application of low-frequency RF is the induction range depicted in Fig. 1-24. Here, the inductor, L, is formed as a flat spiral so that optimum coupling can be had with the cooking utensil. Copper vessels are best for this purpose, but aluminum and iron pots can also be accommodated. This basic scheme has also been used to exploit the heating effect of magnetic hysteresis, as well as eddy currents. Some of the stainless-steel alloys might be good candidates for such heat generation. Control of the "burner" is generally brought about by means of a variable power supply.

A push-pull SCR inverter is shown in Fig. 1-25. The gates are triggered by pulses displaced by 180 electrical degrees. The components shown enable a load power of 1 kW at 13.5 kHz. The basic circuit is very suitable for frequencies up to the highest permissible from readily available SCRs. This limit has been in the vicinity of 25 or 30 kHz for several years, but 50 kHz might be realizable at reduced power levels from some of the latest high-frequency inverter-type SCRs. The push-pull configuration

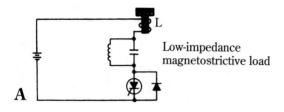

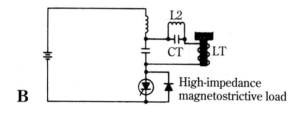

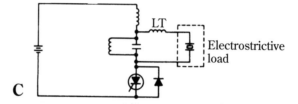

1-23 Methods of coupling ultrasonic transducers to SCR inverter. A. Transducer winding is inductor, L. DC bias is provided for transducer. B. Capacitor C_T series-resonates L_T. L_2 enables dc bias for transducer winding. C. L_T series-resonates capacitive reactance of transducer.

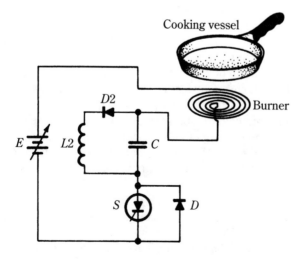

1-24 Basic arrangement for induction range. With L in the form of a flat spiral, the cooking utensil is heated by induced eddy currents.

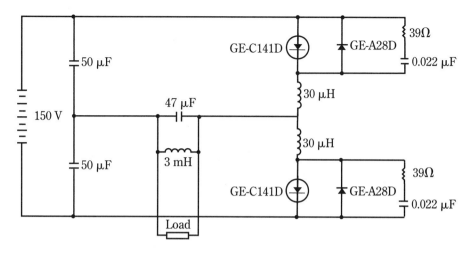

1-25 Push-pull inverter (13.5 kHz, 1000 W). Sinusoidal wave form is more readily developed by push-pull and bridge configurations than with single SCR circuits. _{General Electric Co.}

develops a closer approach to sinusoidal output than the single-SCR inverter. Because it does this at lower triggering rates, commutation difficulties at the higher frequencies are more readily overcome. Although the power supply in Fig. 1-25 is ac center tapped via the 50 microfarad (µF) capacitors, an even better technique is to use a center-tapped power supply, that is, two series-connected 75-V supplies. The circuit can be made short-circuit proof and current limiting by inserting a 1-µF capacitor in series with the load. The component values can be readily scaled to comply with other than the stipulated frequency.

Low-frequency solid-state transmitters

Figure 1-26 shows the Continental Electronics type 314E (AN/FRT-89) low-frequency solid-state transmitter. This equipment has a 2-kW power output rating and is intended for operation in the 275- to 530-kHz frequency range. Some of its applications are as follows:

- Ship-to-shore and shore-to-ship communications
- 500-kHz emergency traffic
- Nondirectional beacon (NDB) navigation-aid transmitter for civil-aviation use
- Meteorology information station
- Geophysical and oceanology communication bases
- Automated merchant-vessel reporting program (AMVER)
- Replacement for tube transmitters—especially where greater efficiency and higher reliability are important factors

The block diagram of this transmitter, depicted in Fig. 1-27, is interesting in that

1-26 A 2-kW solid-state transmitter for 275- 530-kHz operation.

the push-pull power amplifier is driven from a square-wave source. Such class-D operation enables the power amplifier to achieve exceptionally high efficiency by virtue of its essential behavior as a switch. Moreover, at even lower frequencies, very similar designs simply substitute SCRs for the power transistors. Thus, below approximately 30 kHz, and where many kilowatts of output power are needed, you would expect to find SCRs. Whether SCRs or transistors are used as switching devices, the basic idea is that an ideal switch is 100% efficient. The actual efficiency of solid-state

Low-frequency applications 33

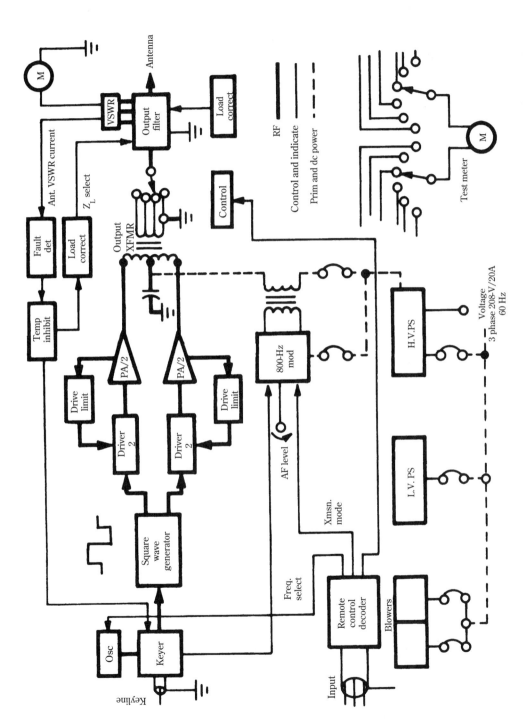

1-27 Block diagram of 2-kW solid-state low-frequency transmitter. The salient feature is class-D operation of the push-pull amplifier.
Continental Electronics

switching devices in such low-frequency transmitters is often well above 80%, and with SCRs at very low frequencies, it can be around 90%. This is the overall efficiency of the final output stage, excluding the harmonic filter.

Radio amateurs and others familiar with class-C and class-B final amplifiers are aware that these operational modes, too, are described as switching excursions. However, in these instances, the transitions from conducting to nonconducting states, and vice versa, are much slower than in the class-D mode. Accordingly, the switching losses are considerably greater. Also, power transistors are much more vulnerable to destruction from second-breakdown phenomena when the switching rise and fall times are slow.

Harmonic energy composes a considerable portion of the square-wave output from a class-D power amplifier. Although most of such energy is not dissipated as heat losses in an output harmonic filter, it nonetheless reduces the overall operating efficiency of the transmitter. This is true because energy was consumed from the dc supplies in order to generate and process the square waveform in the first place. The 2-kW low-frequency transmitter utilizes an interesting method for recovering much of this otherwise unused harmonic energy. As shown in Fig. 1-28, a band-pass filter is used to extract the fundamental frequency for delivery to the antenna system. A band-reject filter channels the harmonic energy to a rectifier, and the dc power thereby produced is used within the transmitter itself. A similar scheme can be de-

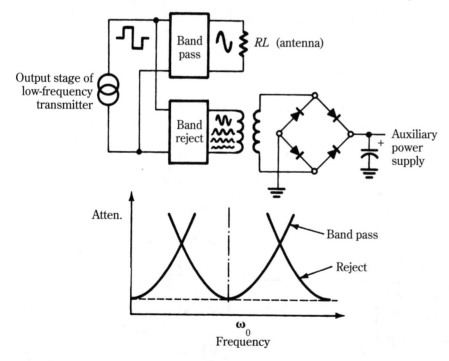

1-28 Recapture of harmonic energy in square-wave output of low-frequency class-D power amplifier. The harmonic energy is rectified and filtered to provide auxiliary dc operating power for the transmitter.

vised with low-pass and high-pass filters. In any event, it is much easier to attain good filter performance at low RF frequencies: inductors and capacitors behave in more ideal fashion and less trouble is encountered with harmonics and spurious frequencies bypassing the filter networks. For example, in the 2-kW solid-state transmitter described, harmonic radiation is on the order of 70 dB below full-power output into the antenna.

The phenomenon of skin effect

The practical manifestation of the current-distribution phenomenon that is known as *skin effect* is that alternating currents concentrate increasingly near the surface of a conductor. At dc, the current density is uniform throughout the cross-section of the conductor. The effective resistance of a conductor is greater at ac than at dc. Indeed, the higher the frequency, the greater is the disparity between dc and ac resistance. A simplistic, but entirely practical explanation is that magnetic flux-linkages are more concentrated at the interior than at the exterior of a conductor that carries ac; consequently, the interior regions present more inductive reactance to current flow than do regions closer to the surface of the conductor. Figure 1-29 illustrates the dependency of skin effect on frequency.

The following statements should remove several common preconceived notions about skin effect:

- Skin effect considerations pertain to the entire spectrum of alternating currents. It is not just an RF phenomenon; power engineers who design ac machinery and ac transmission lines, must allow for skin effect with regard to such matters as safe current-carrying capacity, temperature rise, and efficiency. Thus, though relatively small at 60-Hz, skin effect is not always negligible.
- Skin depth, δ, is that level beneath the surface of a conductor where current density is $1/e$ (approximately 37% of its surface value). Thus, current density does not abruptly become zero at distance δ, but continues to diminish as regions of the conductor closer to the center are approached (Fig. 1-30).

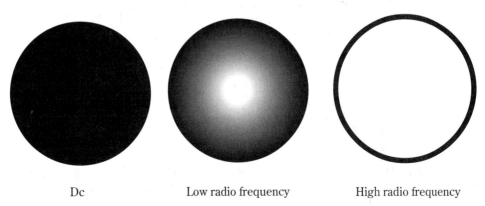

Dc   Low radio frequency   High radio frequency

1-29 The distribution of current density in a round conductor.

36 Transistors versus tubes

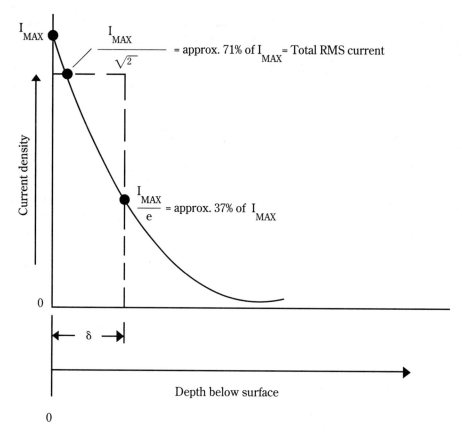

1-30 A graphic depiction of skin depth, δ. A ficticious conductor having a thickness equal to δ and a uniform current density of 71% of I_{max} would have the same resistance as the conductor that is represented by the exponential curve.

- Although it is tempting to suppose that skin effect is caused by the same principle responsible for radiation, this is not so. For example, the radiation resistance of a half-wavelength dipole antenna in free space is 73 Ω at both very low and very high frequencies, despite the great differences in skin effect at these disparate frequencies. Although skin effect crowds current near the surface of a conductor, it, of itself, does not impel the electrical energy to escape from the confines of the conductor. It is only true that RF antenna currents are near the surface prior to radiation. Notice that radiation from a tank inductor is relatively low, whereas it is relatively high from an antenna—even though both might exhibit the same skin effect.

The formula for calculating the skin depth, δ, follows. Recall that δ is the penetration depth beneath the surface of the conductor at which current density is $1/e$ (37% of the current density at the surface).

$$\delta = \sqrt{\frac{\rho}{\pi \times \mu \times f}} \text{ in meters.}$$

where:

ρ is the conductivity of the conductor in ohm-meters (Ω/M) = 1.724×10^{-8} Ω/M for copper (the mks system of units is favored in physics texts).

μ is the permeability of the conductor. Although μ is normally thought of as being one for nonmagnetic substances, that assumption stems from the Gaussian system of units commonly used in engineering textbooks. The mks value of μ is 1.26×10^{-6} henry per meter (H/M).

f is the frequency in Hertz. The losses caused by current concentration near the surface are the same as if the total RMS current was of uniform density to the depth, δ.

For copper conductors, this formula can be simplified to

$$\delta = \frac{2.60}{\sqrt{f}} \text{ in inches.}$$

Also,

$$\delta = \frac{6.60}{\sqrt{f}} \text{ in centimeters.}$$

where:

f remains the frequency in Hertz.

From these equations, it can be quickly calculated that δ = 0.0026 inch at 1 MHz, 0.0082 inch at 10 MHz, and 0.00026 inch at 100 MHz. At the microwave frequency of 3000 MHz, δ is only 0.00005 inch. It can be appreciated that at RF, most of the conductor cross-section contributes more to mechanical integrity, rather than to electrical conductivity! This, of course, accounts for the use of hollow tubings for tank-circuit inductors. Also, recalling that silver is a slightly-better conductor than copper, it can be seen why silver plating on the surface of a copper inductor can increase the Q of the resonant tank circuit. Silver plating is even more profitable when it is applied to brass surfaces, as often is the case in microwave RF hardware.

On the other hand, treating the surface of an RF inductor with paint or varnish does not significantly increase losses. This is because of the very high resistivity of such protective substances; the higher the resistivity, the greater is δ, and an inordinately thick "layer" of a protective agent would be needed to contribute to circuit losses. Oxidation or corrosion that can pit the surface of a copper conductor, can, however, introduce losses and degrade the Q.

The concept of using hollow tubing to provide more surface area for RF currents can be applied in another way. Litz wire has long been available to increase the Q of RF resonant circuits. Litz wire is mechanically flexible because it is actually stranded wire. Unlike conventional stranded-wire, each individual wire is insulated;

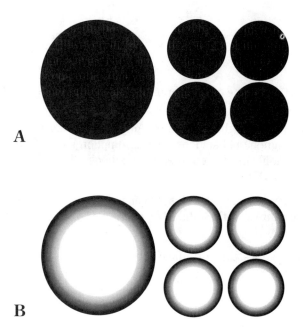

1-31 Current distribution in a single wire and in Litz-wire array of the same area. A. For dc, both conductor systems exhibit equal current-carrying capability and present the same resistance. B. For RF, more conductor area is available for current in the Litz-wire array than in the single wire. Accordingly, RF resistance is less for the Litz-wire array than for the single wire.

interwire connections are made only at the ends of the coil or other elements of the RF circuit. Considerable twisting is imparted by the manufacturer so that each current-carrying wire undergoes frequent transpositions. It is interesting that although the dc current-carrying capacity of Litz wire is less than equivalent-gauge solid wire, the RF resistance can be significantly smaller. This is because the effective surface area of the multiple strands exceeds that of equivalent-gauge solid wire (Fig. 1-31) "Equivalent gauge" infers the same or nearly the same number of ohms per thousand feet.

Many types of Litz wire are made in order to optimize results at different regions of the RF spectrum. Additionally, the gauge and number of strands of the stranded wires must satisfy the power requirements of the intended use. For receivers, Litz wire resembles various gauges of ordinary flexible hookup wire. For transmitters, Litz cables have been fabricated as large as 3.5 inches in diameter (Fig. 1-32). A universal feature of Litz wire is high cost; designers usually try to circumvent its application by using a large diameter of solid wire consistent with other factors, or by attaining selectivity or harmonic suppression via other means.

It is to be emphasized that the low RF loss of Litz wire depends on the insulation of the individual wires; ordinary stranded wire in which individual wires are in mutual contact will not exhibit the same results. An interesting practical observation is that Litz wire performance is not significantly degraded—even when there is considerable internal breakage of the individual wires.

1-32 Low-frequency high-power tuning components wound with 3½-inch Litz "wire."

Although a small gauge of Litz wire might yield high Q in a resonant circuit over an appreciable frequency range, inferior performance at lower frequencies can cause unanticipated troubles in some applications. This is usually easily avoided by appropriately selecting the gauge and type of Litz wire; it is only necessary to pay heed to the manufacturer's published data and performance graphs. Litz wire coils have been mostly used at lower and moderate radio frequencies; above 2 MHz or so, eddy currents and dielectric losses negate the available benefits from this conductor format. Switchmode power supplies that operate in the 20- to 100-kHz range often use transformers and inductors that are wound with Litz wire. Such power supplies rely heavily on RF techniques. A partial list of Litz wire formats is shown in Table (1-1).

A practical tip for winding toroidal inductors

On the one hand, RF toroidal inductors and transformers are easy to handwind because relatively few turns are required, compared to their low-frequency counterparts. On the other hand, it is not always easy to obtain optimum performance, despite the much-heralded attributes of the toroidal configuration. Usually, the two enemies of desired operation turn out to be leakage inductance and capacitance. To a considerable extent, leakage inductance can be minimized by distributing the wind-

Table 1-2. A partial list of Litz-wire formats.

Number of strands	Size (AWG)	Circular mils nominal	Nearest AWG equiv. (cir. mils)	Resistance (ohms per 1000 ft. at 20°C) nominal	Single polyurethane mean o.d. (inches) nominal	feet per pound	pounds per 1000 ft.
3	30	300.00	25½	34.57	.022	1068	.936
4	30	400.00	24	25.93	.025	801	1.25
5	30	500.00	23	20.74	.028	641	1.56
6	30	600.00	22½	17.29	.031	534	1.87
7	30	700.00	21½	14.82	.033	458	2.18
8	30	800.00	21	12.96	.036	401	2.49
9	30	900.00	20½	11.52	.038	356	2.81
10	30	1000.00	20	10.37	.040	321	3.12
15	30	1500.00	18½	6.91	.049	214	4.67
20	30	2000.00	17	5.19	.056	160	6.25
3	32	192.00	27	54.00	.018	1695	.590
4	32	256.00	26	40.50	.020	1272	.786
5	32	320.00	25	32.40	.023	1017	.983
6	32	384.00	24½	27.00	.025	848	1.18
7	32	448.00	23½	23.14	.027	727	1.38
8	32	512.00	23	20.25	.029	636	1.57
9	32	576.00	22½	18.00	.031	565	1.77
10	32	640.00	22	16.20	.032	509	1.97
15	32	960.00	20½	10.80	.039	339	2.95
20	32	1280.00	19	8.10	.046	254	3.94
3	34	119.07	29½	87.10	.014	2680	.373
4	34	158.76	28	65.33	.016	2010	.498
5	34	198.45	27	52.26	.018	1608	.622
6	34	238.14	26½	43.55	.020	1340	.746
7	34	277.83	25½	37.33	.021	1148	.871
8	34	317.52	25	32.66	.023	1005	.995
9	34	357.21	24½	29.03	.024	893	1.12
10	34	396.90	24	26.13	.026	804	1.24
15	34	595.35	22½	17.42	.031	536	1.87
20	34	793.80	21	13.07	.036	402	2.49
3	36	75.00	31	138.27	.011	4230	.236
4	36	100.00	30	103.70	.013	3173	.315
5	36	125.00	29	82.96	.015	2538	.394
6	36	150.00	28	69.13	.016	2115	.473
7	36	175.00	27½	59.26	.017	1813	.552
8	36	200.00	27	51.85	.018	1586	.631
9	36	225.00	26½	46.09	.019	1410	.709
10	36	250.00	26	41.48	.021	1269	.788
15	36	375.00	24½	27.65	.025	846	1.18
20	36	500.00	23	20.74	.029	635	1.58
25	36	625.00	22	16.59	.032	508	1.97
30	36	750.00	21½	13.83	.035	423	2.36
40	36	1000.00	20	10.37	.041	317	3.16
50	36	1250.00	19	8.30	.046	254	3.94
60	36	1500.00	18½	6.91	.050	212	4.72

MWS Wire Industries

The phenomenon of skin effect 41

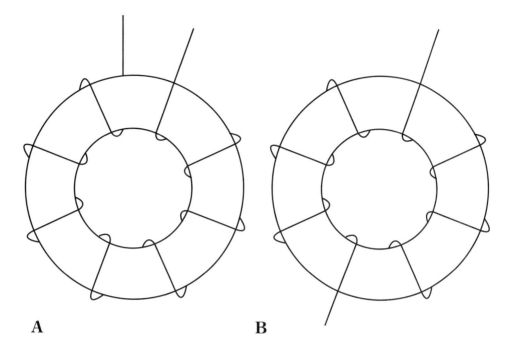

1-33 Two winding techniques for toroidal RF inductors or transformers. A. Conventional distributed winding. B. Super toroid winding.

ing over the major portion of the core's circumference (Fig. 1-33A). This technique also serves to decrease the distributed capacitance prevailing between turns, thereby reducing undesired resonance effects.

For some purposes, the fact that the two leads are brought in close can pose problems. For what has been gained in reducing inter-turn distributed capacity can be largely negated by the capacitive effects between the two leads. This is because the highest difference in RF potential exists at the beginning and end of the winding, and a little capacitance goes a long way in lowering the self-resonant frequency of the inductor. As in RF chokes, both series and parallel resonances can appear; this sometimes accounts for "mysterious" behavior of inductors and transformers. Yet another shortcoming of such windings is the tendency for arcing at the leads when high power, or sometimes moderate power and high Q are involved.

The so-called super-toroid winding of Fig. 1-33B overcomes the possible detriments associated with the conventional winding technique. Here, the winding is split into two equal-turn sections and applied to the two halves of the core. In order that the two windings do not neutralize each other's inductance, the winding direction must be opposite on each side of the core. Notice that the leads, instead of being physically adjacent (Fig. 1-33A) are now diametrically spaced from one another.

Although simple inductors are shown in Fig. 1-33, both winding techniques can be used in conventional transformers, transmission-line transformers, and in baluns. Not only are the two winding techniques amenable to addition of other windings, but bifilar and multifilar windings can also be used. Often one or other of the two winding methods will be preferable because of convenience of lead interconnections. The use of the super-toroid winding techniques seems to confer greater immunity to stray-field pickup than the conventional winding. To the extent this is true, this winding method can also be expected to cause reduced stray fields in tight compartments.

Simple RF power measurements and some simple pitfalls

A commonly-desired RF measurement is the power a transmitter or an RF source is capable of delivering into a 50-Ω dummy load. The salient features of such a load is that it is predominantly resistive in nature, and that it can safely dissipate a known maximum power at least for a time interval that is sufficient to make a measurement in actual practice. In most instances, the desired power measurement will be premised upon sinusoidal RF voltage and current, and this will be the assumption in the ensuing section.

Because the dummy load is resistive, the power factor does not need to enter the picture. The objective is to determine either the RF current consumed by the load, or the RF voltage applied across it. From electrical engineering, the power that can be thereby calculated is simply E^2R or I^2R. In these ohm's-law relationships, the ac voltage, E, and the ac current, I, are effective (RMS) values; this is a nice situation because ac voltmeters and ac ammeters are generally calibrated to read these values. Notice, however, $E \times I$. The product of effective values of sine-wave voltage and current yields what is generally referred to as "average" power. Would not effective power be the goal of your measurement?

It turns out that average power and effective power are identical in the previously described situation, so no harm is done with this terminology. However, it is possible for some confusion to arise because average voltage is not the same as effective voltage for sine waves, nor is average current the same as effective current. This leads to the apparent paradox that multiplying average voltage by average current will not produce the "average" power measurement that you want. Reiterating, the power levels commonly referred to as "average" power are the same as "effective" power in sine-wave systems. Moreover, such average or effective power produces the same heating effect as would dc voltage and current of the same values as the effective ac voltage and current. Refer to Fig. 1-34 to help remember some of these matters. Notice that maximum and peak values are simply interchangeable words.

Because most voltage and current instruments are calibrated in effective values, rather than average values, the power calculated by such measurements is the power that you are seeking. However, in RF practice, it is not uncommon to measure the peak voltage of the sine waves. Now, a little care is needed to calculate power if you are not to go astray. Remember that your objective is to find the average load power from the measurement of the peak voltage that is developed across the load.

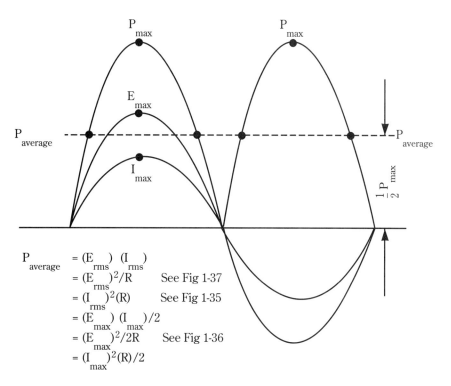

1-34 Ac power relationships in a resistive circuit. Notice that power is always positive and that the power curve is twice the frequency of the voltage and current waves. Also, contrary to common sense, average power is *not* the product of average voltage and average current.

If you calculate power from:

$$\frac{(E_{\text{peak}})^2}{R}$$

the result in peak power. This is often a useful quantity in both design and operation. Very conveniently, the usually sought average power is just one-half of the peak-power value for sine waves. In other words,

$$\frac{(E_{\text{peak}})^2}{R}$$

but in practice you wouldn't commonly use peak-current instrumentation.

For the same reason, the relationship:

$$\frac{(E_{\text{peak}})^2}{2R}$$

in practical RF measurement procedures. However, it provides useful insight and could conceivably be adapted to oscilloscopic techniques, where it is convenient to read peak values.

A direct and simple means of RF power determination is to insert an RF amme-

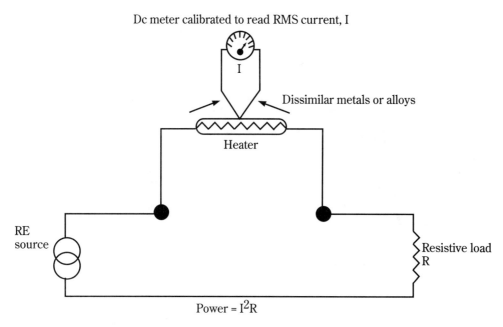

1-35 Use of thermocouple RF ammeter to determine power in a resistive load. Thermocouple ammeter should be accompanied by frequency-dependency data.

ter in series with the dummy load. The long-used instrument for measuring RF current is the thermocouple-type ammeter (Fig. 1-35). Here, a resistive element is heated by the RF current and the thermocouple develops a dc voltage as a function of its temperature. This voltage then produces the deflection of the D'Arsonval meter movement. The scale calibration is conveniently RMS current. Average power is then I^2/R.

A possible pitfall to this method is the frequency dependency to the RF ammeter. As the frequency is increased, skin-effect and stray reactances cause erroneous readings. These effects can manifest themselves seriously at several tens of MHz, or even at lower frequencies. Much depends on the meter construction and its implementation. In practice, the most detrimental effect is from stray capacitance. A meter package in a bakelite or plastic container tends to be better for high frequencies than one that uses metal. In either case, the meter should not be mounted too close to its metal panel or metal enclosure. In the event that reactance is associated with load, R.

Figure 1-36 is the basic circuit arrangement for measuring RF power into a self-contained 50-Ω dummy load. The basic idea is to trap E_{max}, the peak of the sine wave, in the shunt capacitor so that this voltage can be monitored by the high-impedance voltmeter comprised of the sensitive current-meter and its series-multiplying resistor. As shown, the scheme can permit power measurements up to 17-W, or so. The dummy load must, of course be capable of safely dissipating this power level. Two diodes in series would enable extension of the power capability by a factor of four (about 70 W), again with an appropriately selected 50-Ω dummy load. For higher power ranges than this, it would be best to use a voltage-divider network connected across the 50-Ω dummy load.

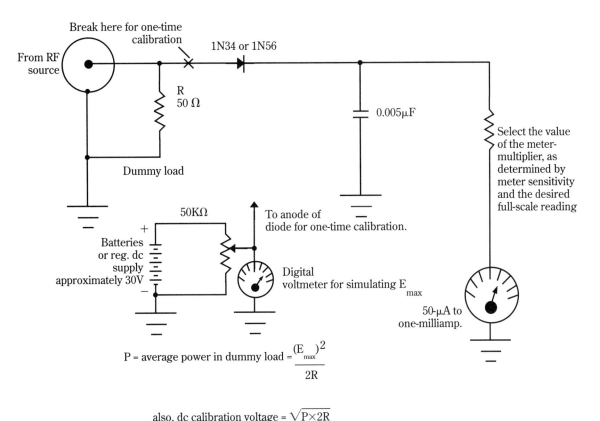

1-36 A basic peak-responding circuit for measuring RF power in dummy load.

For one-time calibration, the measurement circuit is severed at point X and a precise series of dc voltages are injected directly into the anode of the diode. These calibration voltages simulate the E_{max} or peak values of the RF sine wave normally being measured. From:

$$p = \frac{(I_{peak})^2(R)}{2}$$

a table of power calibration points might be prepared. The meter face can then be calibrated to read either peak voltage (E_{max}), or actual power. The series-multiplying resistor for the meter can be experimentally selected to permit the desired meter deflection for a given power level.

Another scheme for RF power measurement is shown in Fig. 1-37. Here, too, the sensed entity is E_{max} (peak voltage), but it is accomplished in a somewhat different way. The capacitor, C, blocks dc, but does not otherwise become involved in the peak-voltage sensing process. Indeed, it could be omitted in situations where it is definitely known that no dc is present. The shunt-connected diode shorts out the positive excursions of the RF sine wave, leaving only negative-going half cycles. The peak voltage values of

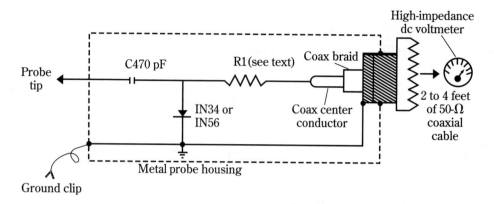

1-37 Basic arrangement for RMS-voltage RF probe. When used in conjunction with a high-impedance dc voltmeter (such as a VTVM, FETVM, or a DVM), it is convenient to measure rms voltage across a known resistance and thereby determine RF power in the resistance.

these negative cycles are then trapped, not in a physical capacitor, but in the capacitance of the length of the coaxial cable, which connects to the VTVM, FETVM, or DVM. Ordinary VOMs do not have sufficient input impedance for use in this scheme.

This circuit is generally constructed in the form of a probe, enclosed in metal shielding. It is very flexible because it is not encumbered with a self-contained dummy load. It can be used to measure power in any known resistance. As will be shown, it uses the existant scales on high-impedance meters.

The value of R_1 is determined by the input resistance of the meter. Meters that are suitable for use with this probe usually have either a 10-MΩ, a 11-MΩ, or a 20-MΩ internal resistance. This calls for R_1 values of 4.1 MΩ, 4.57 MΩ, or 8.28 MΩ, respectively. The basic idea is that R1, together with the meter's input resistance, form a voltage-dividing network so that the voltage delivered to the meter is 0.707 times the available peak voltage. This, in turn, enables the meter to indicate the RMS value of the RF voltage that is sensed by the probe tip. That is, the dc readout of the meter can be read as the RMS value of the sampled RF. The power dissipated in the sampled resistance, R, is then

$$\frac{(E_{max})^2}{2R}$$

or what has been shown to be the same thing,

$$\frac{(E_{RMS})^2}{R}.$$

2

The bipolar transistor in RF power applications

THIS CHAPTER DEALS WITH THE UNIQUE FEATURES OF BIPOLAR RF POWER transistors and their circuitry requirements. This field is specialized, not merely a frequency extension of previous solid-state power systems. Emphasis is placed on the selection of use-optimized (rather than "universal") transistors. As opposed to the first chapter, this chapter emphasizes the differences between tube and transistor approaches to RF power.

What is unique about RF power transistors?

For a long time, the designer and experimenter working with RF power oscillators and amplifiers selected power transistors that had as high a frequency response as was consistent with other parameters, such as voltage and current capability. Generally, this was accomplished by noting the value of f_T in the specification sheet. f_T, the gain-bandwidth product, is the frequency at which the common-emitter current gain has fallen to unity. Naturally, a power transistor that has f_T much higher than the operating frequency is likely to comply with at least one requisite for RF performance. Later, when special switching-type power transistors were developed, you could use the risetimes and falltimes as criteria of satisfactory RF service; a fast "switcher" could be relied upon to perform better at high radio frequencies than a sluggish one.

Although it was evident that many tens of watts could be handled at least to the vicinity of several hundred megahertz, the design and operation of a transistor RF power amplifier remained as much an art as a science. A good measure of success

48 *The bipolar transistor in RF power applications*

was achieved if caution was exercised during tune-up and if the VSWR of the antenna stayed low. In contrast to tube equipment, there was little margin for even momentary deviation from optimum tuning and adjustment. The counterpart of the benign glow of a tube's anode was invariably a blown power transistor. Also, the internal feedback capacitance of these transistors, being high, made some form of neutralization mandatory. This was not objectional per se, but inadvertent oscillation often was sufficient to blow the transistor. Such neutralization tended to be critical and often unreliable.

When ordinary and switching power transistors began to be characterized by safe-operating area curves, it was thought that RF operation, too, could thereby be made safe. Although complying with the operating boundaries stipulated by these graphs was an improvement, it soon became evident that additional "magical" derating had to be applied for RF service. Obviously, what was needed was a power transistor that was specifically designed for RF service. This was forthcoming in the form of a device in which special attention was focused on reducing internal capacitances, preventing hot-spotting (a catastrophic punch-through mechanism that is similar to secondary breakdown, but accompanying RF operation peculiarly), and special consideration for thermal paths. Also, certain tradeoffs were found beneficial. For example, the low collector saturation voltage of the switching transistor could be traded for extended voltage rating.

A unique aspect of bipolar RF power transistors is that operation usually occurs in the attenuation region of the common-emitter current-gain curve. This is shown in a qualitative way in Fig. 2-1. You can see why such transistor amplifiers are prone

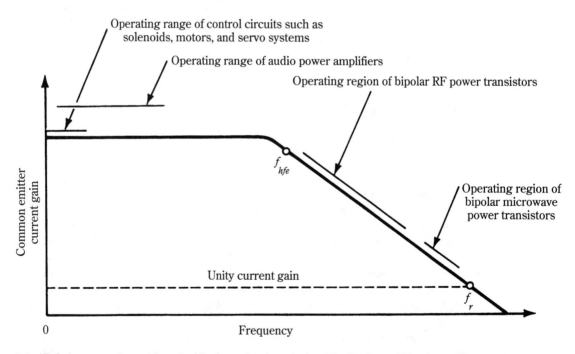

2-1 Relative operating regions for bipolar power transistors. Ideally, it would be desirable to operate RF and microwave transistors in the flat region of the gain curve. Practically, this is not feasible in the frequency range, where most RF power systems operate.

to low-frequency parasitics, oscillations, and instabilities at frequencies lower than the operating frequency. In light of this, it is natural to consider common-base circuits. Indeed, this configuration, with its extended flat-gain characteristic, is frequently used. However, it has its own shortcomings as well. The very low input impedance and very high output impedance often make for awkward impedance-matching problems. Moreover, the common-base circuit is also vulnerable to instability and oscillation at the operating frequency. RF transistors tend to have very low inverse feedback capacitance from collector to base. This enables their use in unneutralized common-emitter circuits. However, a very small stray capacitance between collector and emitter can cause oscillation in the common-base connection. As might be gleaned from these facts, the common-base configuration is often encountered in power oscillators, particularly in the UHF and microwave regions.

RF transistors are designed and packaged to have a very low emitter lead inductance. This is yet another reason why these transistors can be successfully used as common-emitter amplifiers without neutralization. Other sophistications are incorporated in RF power transistors. For example, the inductance, resistance, and capacitance of the base terminal or lead can be manipulated to behave as a low-Q L network. This greatly facilitates the design of broadband amplifiers and tends to make input impedance matching less critical. Packaging provisions are also provided for accommodations to stripline tank circuits. This is very important in making UHF applications predictable and reproducible. The tailoring of stray parameters associated with the leads and the package is an art in itself, and can yield profitable returns when operation is at UHF and microwave frequencies. Table 2-1 provides a dramatic illustration of this by comparing the operation of the same chip in four packaging arrangements. Notice that the coaxial package even excels the stripline package. RF transistor packages are shown in Fig. 2-2.

Table 2-1. RF performance with the same transistor chip in different packages.

	f-GHz	P_{in}-W	P_0-W	P.G.-dB	η_{c}(28V)-%
TO-39	1	0.3	1	5	35
HF-19	1	0.3	1.5	7	45
HF-11	1	0.3	2.2	8.6	50
HF-11	2	0.3	1	5	35

RCA

Gain-leveling provisions

When an amplifier is designed for broadband operation or is intended to be used at different frequencies, it is naturally desirable for its power gain to be reasonably constant. This feature might be desirable for the sake of stability. As already shown, RF gain of a bipolar RF power transistor increases fairly rapidly with decreasing frequency. If matters were left alone, a serious problem could occur, with respect to proper drive power. For example, the drive power level that produces optimum output and/or efficiency at 14 MHz could greatly overdrive the transistor at 3.5 MHz. This could result in an unnecessary temperature rise in the power transistor and

50 *The bipolar transistor in RF power applications*

HF-21
Hermetic
ceramic-metal
coaxial package,
large
(JEDEC TO-201AA)

HF-11
Coaxial package, small
(JEDEC TO - 215AA)

HF-28
Hermetic
ceramic-metal
stripline package
grounded emitter
or base

HF-33
Isolated
electrodes

HF-32
Hermetic
Stripline package

JEDEC TO-72

HF-31
Hermetic
ceramic-metal
stripline package
(Studless JEDEC TO-216AA)

HF-19
Hermetic strip-line type
ceramic-to-metal package
with isolated electrodes
(JEDEC TO-216AA)

JEDEC TO-39

JEDEC TO-60

HF-12
Molded silicone-plastic case
(JEDEC TO-217AA)

2-2 Specialized packaging and mounting provisions used in RF transistors. Power gain, stability, and broadband performance are greatly influenced by parasitic inductance and capacitance of the packaging format. The TO-60 package is available with either grounded or isolated emitter lead.

could adversely affect operating bias and linearity. Moreover, in some situations it could provoke problems in the driver stage. Yet another effect is the forcing of the power amplifier into regions of instability.

Two common circuit techniques are used for gain-leveling the power amplifier so that it will display a near-constant power gain over a desired frequency range. One uses a high-pass filter network in the input circuit (Fig. 2-3). Such a two-element filter network, in conjunction with its load resistance, R, can nearly exactly compensate the 6-dB octave slope of the transistor gain characteristic. This is because the filter network, like the "internal low-pass filter" that you can postulate in the transistor, operates in its attenuation region. With both filters thus providing opposite frequency-response slopes, the net effect is a nearly constant amplitude-versus-frequency characteristic at the output of the transistor. The load resistance, R, absorbs surplus power in inverse proportion to the frequency so that less base drive is applied to the power transistor as the frequency is lowered. At the same time, the driver stage is not disturbed because the power demanded from it is relatively constant, regardless of frequency.

Another gain-leveling technique uses LCR circuits in a negative feedback path shunted around the power amplifier. In Fig. 2-24, the gain is stabilized, with respect to frequency, because the negative feedback increases with reduction of frequency. This method is somewhat less wasteful of power than the aforementioned technique. Such a feedback approach can, if properly deployed, be beneficial in discouraging instability—one of the ever-present problems in bipolar transistor RF amplifiers. Notice, however, that gain-leveling is only effective on one side of the series-circuit resonance curve. If necessary, reduction of amplifier gain at frequencies below f_r can be provided by other means, such as capacitor C_{in}.

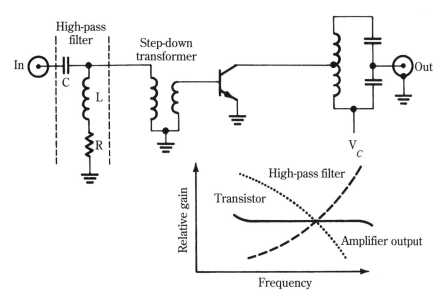

2-3 Gain leveling by input circuit frequency discrimination. The input network, CLR, is a modified high-pass filter with R selected to produce compensation of the transistor's 6-dB/octave roll-off rate. The overall result is a nearly flat frequency response.

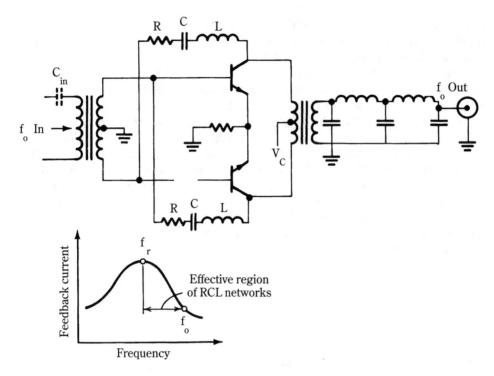

2-4 Gain leveling by means of negative-feedback networks. The RCL networks form low-Q series-resonant circuits, which allow more negative feedback at low than at high frequencies. The effectiveness of this method is limited to the frequency region between f_o and f_r.

Basic transistor class-C amplifier

The circuit shown in Fig. 2-5 is deliberately made similar to that of the basic tube class-C RF amplifier of Fig. 1-5. Fortunately, this involves little compromise; both the tube and transistor operate in much the same manner when performing this function. Nonetheless, certain practical differences will be observed in the transistor circuit. First, the transistor is not a forgiving device. In the face of abusive operation, it will go into thermal runaway, die from secondary breakdown, or be damaged from some other punch-through mechanism. You do not have the option of watching for a glowing anode. Such vulnerability limits freedom of experimentation and even requires caution during ordinary tuning procedures. Moreover, the failure mode is usually of the worst sort: a short between emitter and collector. Thus, an overtaxed transistor amplifier can endanger the power supply, unless suitable protection is provided. The most general failure mode of tubes, diminished emission, usually is of benign consequence.

When a transistor class-C amplifier is operated, the constructor accustomed to the behavior of tube circuits might feel lost, for the time-honored manifestations of RF energy might not be present. Even if 100 W of power is available, its presence might not be indicated by the glow of a neon lamp. Nor will a nice RF arc likely be

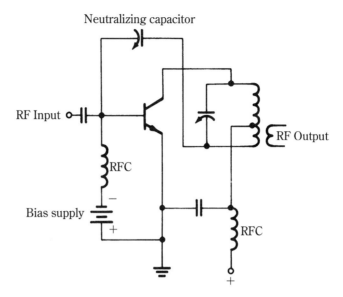

2-5 The basic arrangement for demonstrating transistor class-C amplifier. In practical applications, it is often feasible to dispense with neutralization, and it is not usually necessary to provide fixed bias. Also, one or more of the RF chokes can generally be omitted.

drawn by a pencil touched to a "hot" point (Fig. 2-6). The reason for such apparently passive performance is that the transistor operates at low voltage and high current, rather than the converse situation, which occurs in tubes. However, the RF power can readily be available by conducting suitable measurements at relatively low impedance levels. Although sparking and arcing might be conspicuous by their absence, heating of inductors, capacitors, and leads might be very much in evidence from the relatively high RF currents.

Another difference from tube amplifiers can be observed in push-pull stages. The push-pull tube amplifier often does an effective job of reducing the output of even-order harmonics. Its transistor counterpart is not likely to do so unless a special effort is made to use matched transistors. This is because transistor parameter tolerances are much sloppier than those of tubes.

Although high-frequency parasitic oscillations can plague transistor amplifiers in much the same manner as tubes, circuit instabilities are more likely to occur at relatively low frequencies. This is because transistors are often operated at frequencies where their gain is considerably less than at lower frequencies. Thus, resonances of RF chokes used in the input and output circuits often give rise to mysterious oscillations in transistor class-C amplifiers.

Matching networks

Perhaps it is a surprise to learn that success with solid-state RF power amplifiers is more intimately bound up with the proper use of matching networks, that is, tank circuits, than any other aspect of electronic circuitry. Although skillful design of the LC

54 *The bipolar transistor in RF power applications*

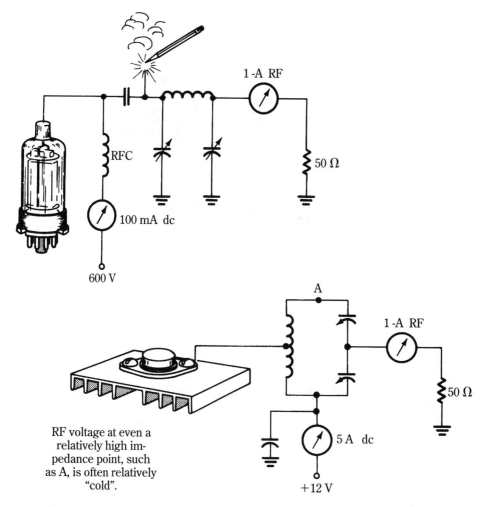

2-6 "Hot" and "cold" RF power. Despite the more spectacular manifestations of RF power in the tube circuit, both RF power amplifiers deliver the same power to the load.

circuits is also important for optimization of tube amplifiers, in practice it is easier to satisfy the requirements of a tube operating at, for example, 600 V and 100 milliamperes (mA) than those of a transistor consuming 3 A from a 20-V source. In the former case, it is likely that a combination of coils and resonating capacitors can be found in the experimenter's junk box, which can produce fair results, or at least point out the path for improvement. However, you are not so likely to come up with a heavy current-carrying inductor and the relatively large capacitors that are suitable for the transistor amplifier. The transistor is not as tolerant of departure from proper design of the resonant tanks as is the tube. One reason for this is that transistors are often operated in the region where current gain is appreciably less than at lower frequencies. With an improper output circuit, the transistor is prone to low-frequency instability, an affliction encountered with some tubes, but with much less severity.

Actually, three important tasks are assigned to the output network or tank. First, energy storage must be sufficient so that a good sine wave is developed at the operating frequency. Otherwise, class-B or class-C operation cannot be obtained. A second and related function is the attenuation of harmonics. This feature is worthy of separate mention because a sine wave that appears good to the eye can still contain appreciable harmonic energy. The third function of the output network is as an impedance transformer so that power can be efficiently coupled from the transistor to the antenna. In practice, a fourth property might be cited, the ability to tune out a small amount of reactance at the antenna feedline.

A number of contradictions, assumptions, and compromises enter the picture. For example, a network should have a high operating Q in order to attenuate harmonics. However, high Q brings with it high circulating currents, which produce intolerable dissipative losses by the time the operating Q is allowed to exceed 20 or so. If the broadband characteristics of the tuned circuit are desired, it is often quite difficult to achieve a sufficiently high operating Q. Further complicating matters is the fact that the operating Q and the impedance transformation values are interdependent. Finally, at VHF and UHF frequencies, the output impedance of transistors varies considerably over the collector voltage cycle.

Parallel-tuned output networks

The simplest output circuit for a transistor RF amplifier is a parallel-resonant LC circuit with either capacitor coupling to the load or inductive coupling via a link (Fig. 2-7A, B, and C). This is reminiscent of similar schemes once extensively used with tube RF amplifiers. The same shortcomings occur with transistors as with tube circuits: harmonic rejection is not very good, and impedance matching to the load (particularly to an antenna feeder) leaves something to be desired.

The output networks (Fig. 2-7D, E, and F) are somewhat better. They provide

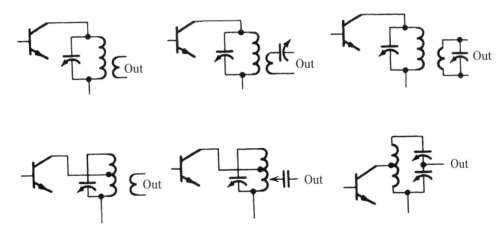

2-7 The various versions of parallel-resonant output tanks that are used with transistors. Networks A, B, and C are best used with amplifiers that operate at power levels up to several watts. At higher powers, networks D, E, and F are likely to provide better impedance matches.

better opportunities for impedance matching and thereby also make better use of the tuned tank for attenuating harmonics. These networks are not, however, as good in these respects as pi and pi-L networks. Also, tee networks are often better performers than parallel resonant tanks. Nonetheless, the parallel-tuned tank circuit is often satisfactory for low power stages and output amplifiers. The network shown in Fig. 2-7F is especially useful in these applications. For frequencies beyond several hundred Megahertz, it becomes increasingly difficult to make effective use of "lumped" LC components, and stripline and other transmission-line techniques are used more frequently.

How transistor and tube differences affect tuned circuits A basic difference between tube and transistor RF amplifiers is in the impedance level of the output circuits. This greatly affects the design of the output tank circuit. To appreciate this, contemplate that if a tube was available as a 50-W RF amplifier, drawing 3 A from a 20-V supply, its output impedance level would be virtually the same as that of a power transistor operating under such conditions. Specifically, both devices in a class-C circuit would look like an RF source with an impedance of $(20)^2 150 \times 2 = 4.0 \, \Omega$.

The significant point here is that, both the hypothetical thermionic device and the semiconductor device behave similarly insofar as concerns the demands made on the output tank circuit. Thus, the extremely low output impedance of transistor RF amplifiers is not mysteriously related to the device's semiconductor material, but is entirely a function of the operating voltage and current. Further confirmation of this stems from the nature of low-power transistor RF amplifiers; they exhibit tube-like output impedances. For example, a transistor RF amplifier that operates at 50 V with a ½-W output looks like a 2500-Ω source. Interestingly, a tube amplifier with 1-kW output power can also present 2500-Ω impedance to its output tank circuit. Thus, the powerful tube amplifier and the flea-power transistor stage might use similar output networks (except, of course, for voltage and current ratings).

Whereas with tubes, it is often sufficient to merely "apply" the RF excitation to the amplifier input (grid or cathode, as the case might be). With transistors, it is usually desirable to use an input-matching network. At higher frequencies, this becomes increasingly important.

Yet another consideration applies uniquely to transistor RF amplifiers. Even though, as previously mentioned, the transistor might not operate at as high of current gain as it could develop at lower frequencies, the transconductance of power transistors is in an entirely different league from that of tubes. Instead of 10,000 microsiemens (μS), which would make for a hot tube, the transistor might operate at several siemens (S) or more. Because of this, it might not be wise to enable the transistor amplifier to develop too much power gain because it will then be difficult to maintain stability. One way to hold down the gain is to use an output network that does not display too much impedance at the operating frequency. Such a restraint often aggravates the already present conflicts between the network parameters.

So much for the disadvantages. Some favorable features pertain to matching networks for transistors. It is often feasible to use ferrite toroidal inductors with transistors. These greatly reduce feedback problems and are conducive to compact packaging. At higher frequencies, stripline tank circuits circumvent awkward design situations in which tuning reactances and stray reactances become comparable. The stripline technique leads to predictable and reproducible performance at frequencies where lump inductors and capacitors are no longer practical. Because transistor RF power is generated at high current and low voltage, less arcing and flashovers will occur. Finally, it is better to have the RF produced in a device of small dimensions so that short connecting leads can be used with the resonating networks. The ability to thus confine RF to where it belongs alleviates at least some problems involving parasitics, stability, and RFI.

Output circuits for medium- and high-power amplifiers

Transistor amplifiers that are designed to deliver more than about 15 W of RF power to the usual 50-Ω antenna feedline generally do not use either the parallel tuned output tanks or pi networks, which have been popular with tube amplifiers. An attempt to use these LC circuits would be frustrated by the inability to achieve an impedance match or by the requirement for impractically large inductors and/or capacitors. Generally, the low output impedance of the power transistor must be stepped up to the nominally 50-Ω antenna feedline impedance. This is just the converse requirement of tube amplifiers, where the tube might typically present an output impedance that ranges from several hundred to several thousand Ω. As previously mentioned, the difference in output impedances between the two devices stems from their different voltage-current formats: tubes operate at high voltages and relatively low currents; the opposite is true for transistors.

You might obtain practical insight into this situation by experimenting with various operating voltages and currents of tubes and transistors. This is facilitated because the approximate formula for output impedance is the same for both devices. It is as follows:

$$Z_L = \frac{E_{dc}^2}{2P_o}$$

where:
   Z_L is output impedance in ohms.
   E_{dc} is the dc voltage applied to the plate or collector.
   P_o = expected output power in watts.

This equation is applicable to class-C amplifiers. An RF power transistor intended to produce 15 W of output power when operated from a 12-V supply would present an output impedance of 4.8 Ω. If 45 W of RF output power was forthcoming from such a stage, the output impedance would be a mere 1.6 Ω. Such transistor power stages, as well as higher-powered ones, generally must use basic output networks (Fig. 2-8A and B). It is true, of course, that the several-hundred-ohm range of output impedance can be obtained at power output levels that exceed 15 W by using RF power transistors that operate at higher than 12 V. Then, you can conceivably use some version of the parallel-tuned output tank or of the pi-network. However, a limit is imposed on how far you can go in this direction by the unfortunate fact that frequency capability and dc operating voltage are trade-offs in the design of power transistors.

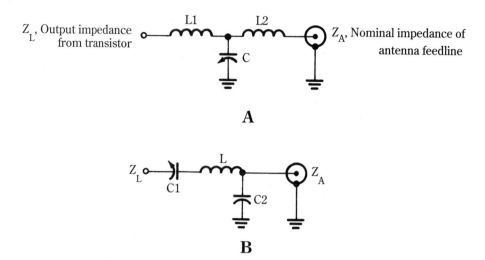

2-8 Output networks that are often used with transistor amplifiers exceeding "Flea-power" levels. A. Tee network. B. L network.

Survival of transistors in RF service

The mysterious demise of power transistors used in switching applications plagued designers a long time before the phenomenon of second breakdown was discovered and properly investigated. From this experience came the *safe-operating area (SOA) curves*. By ascertaining that the dynamic load line of the transistor never penetrated the off-limit boundaries of the SOA, the problem was solved and its "mystery" gave way to a reliable design procedure. Many power transistors that were fast "switchers" also possessed some attributes of a good RF device. In particular, the gain-bandwidth product tended to be high; that is, the frequency response was suitable for at least some RF work. It was found, however, that such transistors were quite vulnerable to catastrophic destruction; this is readily brought about by a departure from optimum tuning or from a moderate or high VSWR caused by an antenna or feeder defect. To make matters worse, there was a pronounced tendency for gradual deterioration—even when operating conditions seemingly were optimized.

Empirically, it was found necessary to operate with a safety factor greater than would be conferred by merely paying heed to the SOA curve. Investigating with an infrared "microscope" revealed that the average temperatures pertaining to the junctions or to the case were the culprits: the concept of average temperature was not valid for RF operation. Instead, there were hot spots in tiny regions within the area of the chip. Because second breakdown, itself, is actually a thermal phenomenon, it can be appreciated that localized pinpoints of higher temperature than that averaged for the entire chip area can manifest themselves in a destructive manner. Although not completely similar, from a practical standpoint it is much as if an earlier occurrence of secondary breakdown occurs. Figure 2-9 shows the practical effect of RF-induced hot-spotting. The allowable safe operating area is curtailed. Fortunately,

Survival of transistors in RF service 59

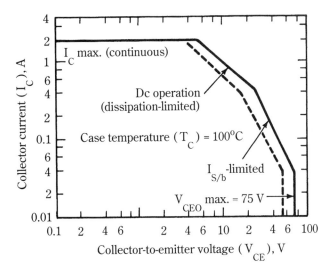

2-9 The general effect of hot-spotting on the SOA. The dashed line shows additional operating limitations that must be imposed on non-RF specified boundaries.

however, it is feasible to construct RF power transistors so that the hot-spotting tendency is greatly diminished.

Techniques to reduce hot-spotting in RF power transistors

The destructive phenomenon of hot-spotting is suggestive of the somewhat more readily identifiable malfunctions that tend to occur when two or more power transistors are connected in parallel in a dc or audio circuit. Even if the transistors are fairly well matched in characteristics, the partitioning of current through them will generally be unequal. Not only does this defeat the basic philosophy of dividing the load power equally among them, but as temperatures rise after initial turn on, the inequality becomes more pronounced, generally with one transistor assuming the major burden of load current. Such current "hogging" is inherently regenerative, because the transistor that begins to hog the current finds itself increasingly provoked to do so as more current is accompanied by increased heat generation. Such cumulative current hogging (thermal runaway) is likely to terminate in the catastrophic destruction of the "hungry" transistor.

The remedy to such malperformance is well known, and is depicted in Fig. 2-10A. It consists of inserting ballast resistances in each emitter connection to the paralleled transistors. The balancing effect of these ballast resistances is straightforward: if any transistor attempts to hog current, the current flowing through its ballast resistance and emitter circuit develops a polarized voltage drop, which decreases the forward bias of its emitter-base circuit. Although various biases will exist among the paralleled transistors, they will incline toward equality of load current division. In most practical applications, you must compromise between the value of the resistances and the quality of current sharing attained. Higher resistance results in better current division, but it also reduces the overall efficiency and voltage-handling capability of the transistor bank. Also, higher resistance decreases power gain because of degeneration.

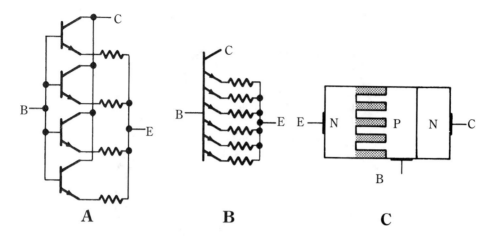

2-10 Techniques for reducing current hogging and hot-spotting in power transistors. The same basic concept applies for all three situations. A. The use of emitter ballast resistances in a discrete transistor circuit. B. A hypothetical structure in which multiple emitters are ballasted. C. An interdigital RF transistor that shows emitter geometry. Shaded emitter "fingers" are processed to have high resistivity.

To provide additional insight into this technique, a hypothetical transistor with multiple emitters is shown in Fig. 2-10B. Here, current hogging by any of the emitters is prevented by the same technique of ballasting. Such a structure is not a fantasy; multiple-emitter transistors are often used in monolithic ICs. Without the ballast resistances, the multiple-emitter transistor would suffer from hot-spotting if the ordinary power transistors were operated at radio frequencies. Although the physics of the two compared situations are not exactly the same, the similarity is, indeed, close enough to enhance understanding.

The RF power transistor illustrated in Fig. 2-10C has emitter geometry with an *interdigited* configuration. Moreover, each of the emitter "fingers" is processed to have higher resistivity than would be the case in ordinary power transistors. This structure then provides the effect of many emitter sites, each with its own ballast resistance. From a circuit viewpoint, these internal resistances are in parallel insofar as their effect on overall degeneration is concerned. Therefore, it is feasible to control the current distribution in the chip without degrading the power gain of the transistor. Because the ratio of emitter periphery to area is high, the emitter section is used effectively.

Bias considerations At first inspection, the base-emitter biasing requirements for bipolar RF transistors seem to pose no unique problems. Class-A stages are not widely used where the objective is the efficient processing of power. The biasing circuits of class-A amplifiers are well known from audio and dc practice. Class-B operation appears to offer no unique aspects, and class-C operation automatically ensues from operating the base at emitter potential. This occurs naturally by grounding the RF choke or resistance in the base-return circuit of grounded-emitter amplifier circuits. If you wish to operate "deeper" into the class-C region, connect the aforementioned choke or resistance to an appropriate negative voltage source, rather than to ground. Why look for nonexistent problems?

Most class-B (linear) amplifiers actually are more accurately described as operating in the class-AB region. That is, they are slightly forward-biased so that a small dc idling current exists in the collector circuit when no RF drive is applied to the amplifier's input. This operational mode pays dividends in the form of considerably improved linearity, compared to "pure" class-B operation. The idling current is adjusted to be a tiny fraction of maximum collector current at ambient temperature. Under this condition, the base-emitter junction of the transistor is approximately at ambient temperature. If RF excitation is applied to the amplifier input for a time, the junctions will undergo temperature rise because of the inescapable power dissipation within the transistor. When the input excitation is removed, the idling current will not return to its initially set small value. This is because the bias voltage that was sufficient to produce the small idling current when the transistor junctions were at, or near, ambient temperature now produces a heavy collector current. This would be serious enough if it merely upset the quiescent operating point of the amplifier.

Because the base-emitter voltage has a negative coefficient of temperature, a situation quickly develops where the collector current develops more heat, and in turn the excessive forward bias of the base-emitter junction begets yet greater collector current. This is, of course, thermal runaway, which seldom satiates itself with less than the catastrophic destruction of the transistor. This phenomenon leaves little doubt that there is, indeed, a biasing problem. Some means is necessary to automatically lower the forward base-emitter bias in response to junction temperature.

A typical method for accomplishing such bias compensation is illustrated by the power amplifier circuit of Fig. 2-11. The principle involved is simple enough: the forward voltage developed across the pn junction of diode D1 decreases with temperature and thereby also causes the forward bias that is applied to the base-emitter circuit of Q1 to decrease. Of course, we would like to achieve exact tracking of the compensation so that the idling current in the collector circuit of Q1 remains constant over a wide temperature range. This objective can be reasonably approached by experimenting with the method of mounting D1 and with resistance R_1. Sometimes a tiny resistance in the ground lead of D1, or a high resistance in parallel with

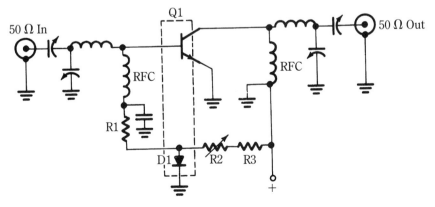

2-11 A basic diode compensation circuit for preventing thermal runaway. The dashed lines indicate close thermal coupling between the RF power transistor and the diode.

62 *The bipolar transistor in RF power applications*

it, will exert useful control effects. If Q1 is a 50-W transistor, R_1 might be 1 Ω or so, and R_2 and R_3 can be 100-Ω resistances with 20-W power ratings. D1 should be a high-conductance silicon diode, such as a 1N4719. For lower-power amplifiers, the 1N4001 is probably more appropriate. Some RF power transistors, such as the RCA 2N6093, contain an integrally packaged compensation diode. This arrangement provides more intimate thermal contact with the transistor junctions and is conducive to better tracking than an externally mounted diode.

A more elegant scheme for compensating the effect of junction temperature is shown in the biasing arrangement of Fig. 2-12. The class-AB amplifier with the RCA 2N6093 RF power transistor develops an output level of 75 W PEP at 30 MHz. The four transistors (Q1 through Q4) might be viewed as a current amplifier in the sense that transistor Q4 is a much more adequate source of biasing current than the simple diode-biasing network of Fig. 2-11. Circuitwise, these four transistors actually compose a voltage-regulated bias source with the reference voltage being derived from the temperature-compensating diode and its resistive network (this diode is integrally packaged with the 2N6093 power transistor). Unlike the typical Zener diode reference, the voltage from the compensation diode is variable. Specifically, it tracks

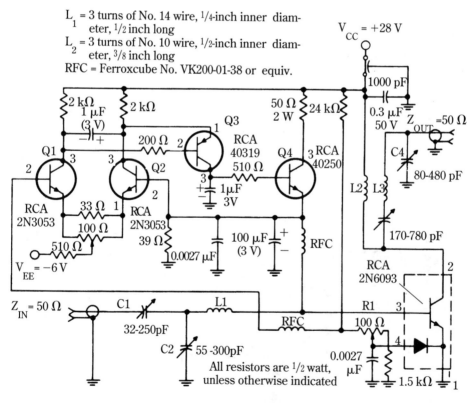

2-12 A bias compensation scheme that uses current amplification. Transistors Q1 through Q4 compose a voltage-regulated bias source that senses the voltage drop across the compensating diode.

the change of the base-emitter voltage of the 2N6093, with respect to temperature. In this way, the bias voltage and current available from the emitter of Q4 is controlled over a wide temperature range to cancel the effect of temperature on the quiescent operating point of the RF power stage.

The effectiveness of this biasing scheme in preventing thermal runaway is shown in Fig. 2-13. What might not be obvious is that the voltage-regulated bias source actually improves the linearity of the amplifier for single-sideband service. This is because such a "stiff" bias source stabilizes the operating point from the effect of the audio-modulated drive signal.

Basic problem of transistor RF power circuits

A two-part problem has plagued transistor power amplifiers and oscillators ever since power transistors became commercially available. The maker has had to contend with the frequency capability of the transistor, and the user has had to become versed in the very exacting art of its implementation. The use of such words as "art" might not seem apropos in the face of much published data pertaining to the characteristics of transistors, sophisticated specification sheets, and precise engineering techniques (such as the use of Smith charts and the availability of high-quality RF instrumentation). Yet, merely being proficient with the mathematical tools needed will not ensure success. This is especially true when working with power levels exceeding several tens of watts and with frequencies of several tens of megahertz and greater. At the microwave frequencies, malperformance is readily obtained—even at very low powers. It is only natural to ponder the common problem in these applications.

A little investigation reveals that the difficulties stem from low impedance and/or high RF currents. Remember that both input and output impedances of tran-

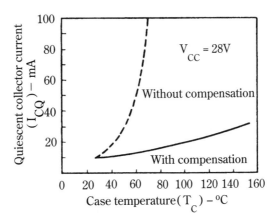

2-13 Effect on thermal runaway of the bias compensation scheme of Fig. 2-12. Without bias compensation, thermal runaway would set in as the case temperature of the 2N6093 approaches 80°C. Linearity and stability would be adversely affected at even lower case temperatures.

sistors decrease with increasing frequency, and most certainly decrease with power level. If, then, you desire both high-frequency operation and a high power level, you must be prepared to use tank circuits and impedance-matching networks with much lower impedances than you might have been accustomed to when working with tube equipment or with low-level solid-state oscillators and amplifiers. Impedance levels of a fraction of an ohm to several tens of ohms are the order of the day in high-power, high-frequency circuits. At these levels, it indeed becomes an art to connect a capacitor to the circuit when the leads of the capacitor have almost as much inductive reactance as the capacitive reactance being "connected" to the circuit. Those very same leads might have sufficient resistance to spoil the calculated Q of the network or even to heat to fusing temperature from the heavy circulating RF current.

While on the subject of capacitors, the "capacity" can be misleading or meaningless unless it is measured at the frequency of use. The reason is not only because of stray inductance in the leads and structure, but because the dielectric material might behave differently at different regions of the RF spectrum. Other mechanisms, such as fringing and radiation, also impart a frequency characteristic to real-world capacitors. Capacitance to ground planes and other components are yet more reasons why the RF circuit might "see" a different capacity than is stamped on the physical capacitor. Adding further to your difficulties is the problem of ascertaining that a network capacitor is truly connected where it is intended to be. Where a ⅛" placement difference on a circuit board would be little cause for concern at several megahertz, such sloppyness could cause appreciable detuning at several hundred megahertz.

The traditional way of selecting input or output coupling capacitors or dc-blocking capacitors (assuming such capacitors are not actually part of a tuned network) is to merely ascertain that the calculated capacitive reactance is "very low" at the operating frequency. Thus, a 0.01-µF ceramic capacitor might be stipulated for a several megahertz circuit working into a 50-Ω load. According to this philosophy, there would be no adverse effects (other than bulk and cost) if the capacitor was made larger, 0.1 µF (for example). However, by the time you get into the several tens of megahertz region, and certainly at several hundreds of megahertz, the naivete of such an approach becomes painfully manifested. What has been neglected is the *intrinsic inductance*, that caused by the structure and leads of the capacitor. This makes every capacitor behave as a series-resonant circuit. If a large capacitor is chosen and it turns out that its effective series resonance is near, but not at, the operating frequency, the circuit will not "see" a capacitive component with low impedance, but rather it will "see" a relatively high impedance path.

From the preceding consideration, a more refined design approach is necessary to select the capacitance no greater than corresponds to series resonance at the frequency of interest. If broadband operation is involved, the frequency of interest is the geometric mean. Thus, if the band extends from, say, 100 to 200 MHz, the mid-band (frequency of interest) is $\sqrt{100 \times 200} = \sqrt{20{,}000}$ (approximately 141 MHz). In pursuing this design approach, it is important that the inductive component of the capacitor first be minimized by appropriately selecting the capacitor type, then by minimizing (or eliminating) lead length. It is always best to use capacitors that have been made especially for use in the RF spectrum where they will operate. Such capacitors might have short ribbon leads and might be ceramic, mica, or porcelain types. Even better are those with no leads, the capacitor "chips." Schematics often feature two paralleled capacitors. This technique minimizes lead inductance and increases RF current capability.

Bypass capacitors generally are brute-force components of relatively high capacity. They are often electrolytic, high-dielectric-constant ceramic, or mylar types. In most instances, even when working with low frequencies, these capacitors must themselves be bypassed by one or more smaller RF-type capacitors. Here, too, rewarding results might accrue by taking advantage of series resonance.

The effect of low-impedance transistor circuits on inductors is to make them vanishingly small: electrically and physically. Inductance is often simulated by transmission-line elements. In particular, stripline elements are amenable to calculation and result in reproducibility in manufacturing. This is because such elements are lengths of copper strips on PC boards. Their geometric dimensions, together with the dielectric constant and thickness of the board sandwich material, govern the network behavior of these elements. Most important, they facilitate transformation from and to low impedances.

RF grounding of the common lead

A severe obstacle in dealing with RF power transistors has been the RF grounding of the common lead: the emitter terminal in common-emitter stages and the base terminal in common-base stages. The presence of a fraction of an ohm impedance in the grounding lead can produce pronounced effects in power gain and stability. This is one reason why RF power transistors have evolved with specialized packaging techniques. For example, the opposed emitter case has not one, but two wide terminations for the emitter connection(s). This minimizes the inductance in the emitter grounding path. You are dealing with just a few nanohenrys here, but the inability to achieve a near-zero impedance grounding path was one of the factors that delayed the progress of transistor RF power. Impedance in the emitter-ground path produces negative or inverse feedback, and impedance in the base-ground path of a common-base stage produces positive feedback. In actual practice, such ground-return impedance produces various erratic results in both amplifier configurations.

In summing up, it is clear that the design and construction of successful transistor RF power circuits is, indeed, an art and a science. It is not to be construed that art of implementation implies that the underlying scientific principles are esoteric or difficult to grasp. Rather, the art composes an arsenal of practical techniques to be skillfully deployed in an endeavor to comply with basic principles. These basic principles are generally well reinforced by adequate measurement and performance data that is supplied by the device manufacturer. On the other hand, some empirical investigation will always be needed because of numerous variables that cannot be accounted for. The purely experimental approach has been left behind, however. It is no longer necessary to find out by patient and persistent experimentation whether an audio power transistor will magically perform in an RF application.

Transistor frequency multipliers

Familiarity with tube circuits for producing frequency multiplication might lead to the conclusion that the transistor version would simply emulate tube circuitry and design philosophy. That is, a class-C amplifier with an output circuit tuned to the desired harmonic would be expected to fulfill this function. This is true, and many frequency multipliers have been implemented in just this way. Sometimes empirical investigation of optimum drive and bias levels has paid worthwhile dividends. However, the approach of merely duplicating tube techniques does not use the bipolar transistor to maximum advantage as a frequency multiplier.

The capacitance variation of the collector-base diode with instantaneous collector voltage is well known. Indeed, this varactor phenomenon is responsible for generating more harmonic and intermodulation distortion than would otherwise be the case. Why not use this to advantage in frequency multipliers? Such a question is particularly tantalizing because varactor-diode frequency multipliers have long been efficient frequency multipliers. By efficient, I am implying a relatively low power loss; the passive nature of varactor diodes precludes the possibility of power gain.

It turns out that combined class-C amplifier action and collector-base varactor action can, indeed, yield frequency multipliers with greater harmonic-producing efficiency than is available from the effects of class-C amplifier action alone. To accomplish this, idler circuits must be provided for the circulation of harmonic currents. This is quite similar to practice in varactor-diode frequency multipliers. Figure 2-14 depicts the general configurations of common-base frequency multipliers. The series-resonant idler circuits optimize the flow of fundamental and harmonic frequency currents through the transistor. This both enhances varactor action and provides the opportunity for heterodyne products that are equal in frequency to the desired output harmonic to be generated. With one exception, the remainder of the LC circuitry composes the conventional impedance-matching networks.

The exception is the parallel-tuned tank circuit appearing in the frequency doubler of Fig. 2-14A. This parallel resonant circuit is tuned to f, the fundamental frequency, and functions as a trap to reduce contamination of the $2f$ output by the fundamental. Notice that a $3f$ idler circuit is not needed in the quadrupler. Circuit anomalies of this nature are often learned by empirical investigation.

The common-emitter frequency tripler (Fig. 2-15) also uses collector-base varactor action and is similar to the common-base circuits. However, the L2/C2 series-resonant tank in the input is necessary to establish a complete path for $3f$ (450 MHz) current. L3, L4, and C4 compose idler circuits for the fundamental frequency (150 MHz) and the second harmonic. The 450-MHz impedance-matching network in the output circuit is made up of C5, L5, C6, L6, and C7. Some experimentation is generally worthwhile in order to obtain optimum performance. Generally, compromise is struck between multiplying efficiency and output wave purity. Notice that the idler circuits also function as wave traps. This being the case, it is desirable that these resonant circuits have as high Q as is practical. Their Qs are, unfortunately, largely governed by the losses encountered in the junctions of the transistor).

Because the RCA overlay transistors display inordinately low losses from varactor operation, these transistors are eminently well suited for such frequency-multiplier service. Some, such as the 2N4012, are actually characterized for this mode of frequency multiplication. This particular transistor can provide 2.5 W at 1000 MHz as a frequency tripler, with a collector efficiency of 25%. An interesting aspect

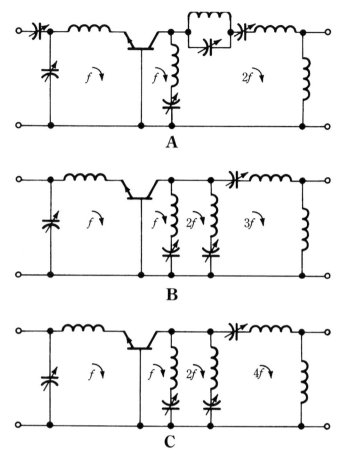

2-14 Common-base frequency multipliers. A. Doubler. B. Tripler. C. Quadrupler. These simplified circuits do not show dc supply and bias provisions.

of exploiting the varactor action of the transistor is that its frequency capability is approximately doubled.

Selecting the RF transistor

The fact that it is of overwhelming importance to select an appropriate device for the particular application at hand is indicative of the progress made in solid-state RF power. Not so long ago, the chief consideration was simply whether amplification or oscillation might occur. You had to be prepared to select from a handful of a given type the particular transistor, which, for a variety of nebulous reasons, yielded the best performance. Although solid-state RF power continues to be both an art and a science, it is no longer a mysterious art. Enough is known, tabulated, and systematized so that each application can be optimally implemented via a standardized

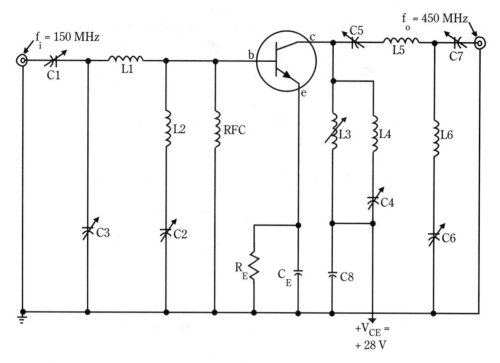

2-15 Common-emitter frequency tripler. This circuit requires a series-resonant idler tank in the input so that $3f$ (450 MHz) can complete its path through the collector-base section of the transistor.

design approach. In selecting appropriate transistors, first recognize that they are manufactured in specific use categories to best satisfy at least one of the following operating features or considerations:

Frequency It is no longer necessary to seek the device with the highest frequency capability, but rather you should pick one that is intended for use in the desired frequency range. Transistors with excessive frequency capability tend to be fragile when used for low-frequency service.

Bandwidth While this performance parameter might initially appear to be governed by the associated LC networks, this is not entirely true, for the stray and distributed reactances and resistances of the transistor package and leads actually constitute part of the tank circuits. For example, some transistors are designed so that their input impedance represents a low Q L-section, which thereby facilitates broadband response.

Power, current, and voltage These parameters are, of course, initially influenced by the circumstances of the application and by cost. Notice that the flexibility is subject to limitations that have to do with inherent trade-offs in semiconductor technology. For example, if you are adamant with regard to frequency and power level, it might be necessary to "choose" an available operating voltage.

Mode of amplifier operation Optimized performance features are designed into transistors that are specifically intended for class-A, class-B, or class-C operation, as well as for linear, AM, video, frequency multiplication, or pulsed service.

Circuit configuration   Transistors are optimally packaged for either common-emitter or common-base operation. Some are also intended for common-collector operation.

General service orientation   RF power transistors might be marketed for amateur, space, military, marine, aircraft, industrial, or other service. This not only influences the basic operating features, such as frequency capability, power, and voltage, but also bears on cost, reliability, and characteristic uniformity.

Package   The packaging of an RF power transistor is no trivial matter. In addition to the requirement of low thermal resistance to a heatsink, these devices have unique relationships to their associated circuits. For example, the seemingly tiny inductance of the emitter lead and its RF grounding path exert an almost controlling effect on power gain and operating stability. Essentially, the same is true for the base lead in common-base circuits. If you intend to use stripline network elements, the appropriate package style is the SOE terminal arrangement (stripline opposed emitter).

Electrical ruggedness   In RF power transistors, this figure of merit applies primarily to the ability of the device to stand up under high (or infinite) VSWR conditions, such as might be incurred by a shorted or open antenna feedline.

Stability   Stable operation greatly depends upon the initial design of the transistor and involves such factors as internal feedback capacitance, lead and packaging parameters, spacing and shielding techniques, power gain, etc. Operational problems can involve both low- and high-frequency parasitic oscillation. Much of the success of producing a good RF power transistor has to do with the ease with which circuit designers can incorporate it in a stable amplifier. The best guide here is to select the transistor supported by the maker's application notes bearing the closest relevance to the intended application. That is, avoid the "general-purpose" approach.

Microwave   Although an extension of the considerations involved under the heading "Frequency" is involved, microwave transistors are unique devices and are often grouped by themselves. In this spectral region, more emphasis is placed on common-base circuits, oscillators, and pulsed amplifiers.

The general ideas incorporated in the preceding discussion of operating features are depicted in the typical listings of RF power transistors (Figs. 2-16 and 2-17).

Avoiding a pitfall

The bulk of transistor specifications pertain to operation under small-signal conditions. The parameters listed are generally suitable for amplifier circuits that are biased for class-A operation. However, most RF power amplifiers operate in class-C, class-B, or class-AB. It turns out that the input and output impedance values of class-A amplifiers can be quite different from those in the other operating modes, which are generally classified as large-signal values. This is why the design and operation of RF power amplifiers long remained a cut-and-try procedure. To achieve engineering predictability, the measurement of impedances must be made under operating conditions that simulate those that are encountered in actual design implementations.

70 The bipolar transistor in RF power applications

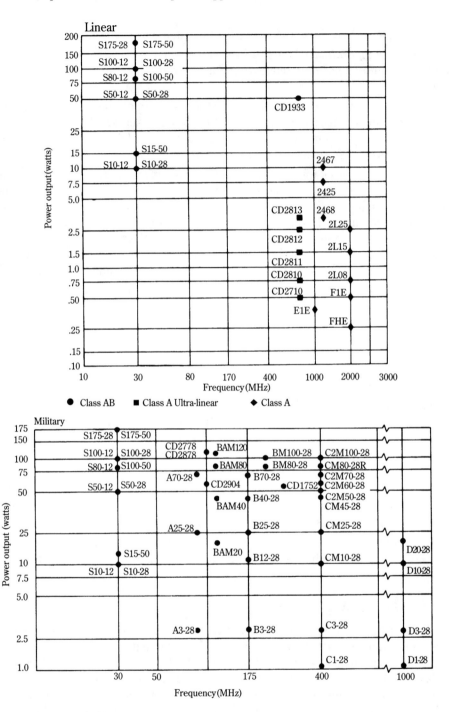

2-16 A typical listing of RF power transistors. Communications Transistor Corp.

Avoiding a pitfall 71

2-16 Continued

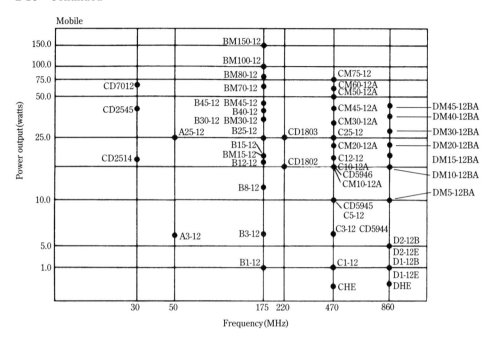

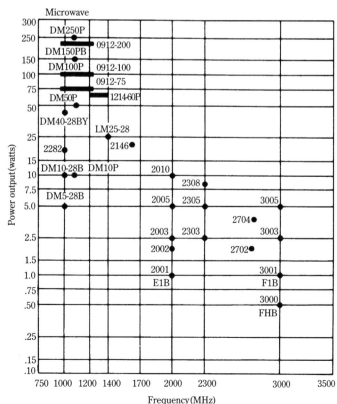

72 The bipolar transistor in RF power applications

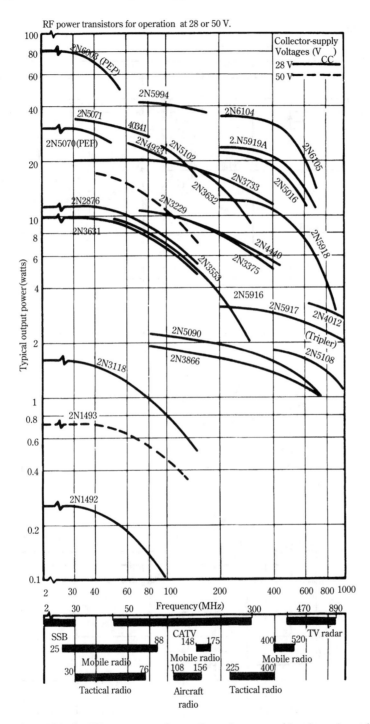

2-17 28- and 50-V bipolar RF power transistors. Impedance-matching is easier at higher dc supply voltages. RCA

It is only natural to ponder the significance of impedance measurements that are derived with large signal swings, such as those that exist in class-C amplifiers. It is clear that any data so recorded can at best be only average values. This is so because a hard-driven transistor behaves as a very nonlinear device. Nonetheless, such impedance values are far more suitable for the design of practical impedance-matching networks than are the small-signal impedance values.

The differences that can occur between small- and large-signal measurement procedures are illustrated in Table 2-2, which compares such impedance data for a 2N3948 transistor. The impedance values depicted are for a parallel circuit. For example, the small-signal input impedance corresponds to a 9-Ω resistance that is shunted by a 0.012-µH inductance. In contrast, the large-signal input impedance corresponds to a 38-Ω resistance, shunted by a 21 pF capacitance. These values can be readily converted into equivalent series-circuit values by means of the conversion equations in chapter 4.

Table 2-2. Small-signal (class A) vs large-signal (class C) parameters.

The data pertain to a 2N3948 transistor and show the divergence between commonly published small-signal information and the parameters relevant to most RF power circuitry.

	Class A Small-signal amplifier V_{CE} =15 Vdc; I_c =80 mA; 300 MHz	Class C Power amplifier V_{CE} =13.6 Vdc; P_0 =1 W.
Input resistance	9 Ω	38 Ω
Input capacitance or inductance	0.012 µH	21 pF
Transistor output resistance	199 Ω	92 Ω
Output capacitance	4.6 pF	5.0 pF
G_{PE}	12.4 dB	8.2 dB

Motorola Semiconductor Products, Inc.

A word is in order with regard to the class-C data. This pertains to the common circuit practice of grounding the base through an RF choke so that both base and emitter are at dc ground potential. As might be suspected, the transistor generally does not operate as deeply into class-C as is often the case with tubes. You might say that such a transistor operates in class BC, although such nomenclature is not common. Generally, the same large-signal data are equally applicable to class-C, class-B, and class-AB amplifiers. The reason is that the class-AB amplifier, as used in linear amplifiers, leans much more to the class-B mode than class A. As you have seen, the class-C amplifier is, itself, probably closer to class B than the traditional class-C operation of tubes. Even where biasing is such that the conduction period is very short (deep class C), as in certain oscillators and frequency multipliers, the same large-signal impedance parameters produce good results.

3
The field-effect transistor in RF power applications

THE TUBELIKE CIRCUIT QUALITIES OF FIELD-EFFECT TRANSISTORS HAVE LONG merited some consideration for RF applications. Up to this time, the power capability of these devices had been so low that their advantages over bipolar transistors were not always compelling. Later development of the power MOSFET (VMOS power FET) radically altered this situation. This chapter develops the fact that these new field-effect devices might represent a major future trend in solid-state RF power. At the very least, these FETs are presently competitive with medium- and high-power bipolar transistors and can offer definite circuit and operational advantages.

Junction field-effect transistor (JFET)

The *junction field-effect transistor* has been a star performer in RF circuits for many years. Unfortunately, their low-power capability has primarily relegated their use to small-signal applications, such as RF amplifiers and local oscillators for receivers. In such service, the JFET has acquired a reputation for low noise, high-frequency capability, small and relatively predictable drift characteristics, and minimal loading of resonant circuits that are connected to its input. Because this book covers solid-state RF power, receiver applications will not be discussed. However, in light of the arbitrary definition of power, some allowance must be made for the somewhat nebulous implication of the term. I look upon current and voltage levels that require a heatsink for the active device to be power circuits. In general, this commences at 1- to 3-W output capability. However, the JFET has seen considerable use in applications where outputs of a fraction of a watt to about one watt have been useful. These applications are still far removed from the several tens of milliwatts (or less) level that is associated with receivers. Moreover, FET functions relate to transmitters. Therefore, these low-power FET applications will be considered.

The low-power FET applications involve oscillators, buffers, frequency multipliers, and final amplifiers in amateur QRP transmitters. Another reason that these small devices merit attention is that the fractional-watt power level they develop at VHF and UHF is often as practical as several tens of watts at lower frequencies.

All JFETs are depletion-mode devices; they produce maximum drain current (or nearly so) at zero gate voltage and require the application of reverse-bias voltage to reduce the drain current (see Fig. 3-1). Both n- and p-channel JFETs are available, sometimes in matched pairs. These are, respectively, similar to npn and pnp bipolar transistors. JFETs are immune to thermal runaway and readily perform well in parallel. However, because parameter tolerances are sloppy (at least in "economy" devices), it is a good idea to select reasonably similar units for paralleling.

JFETs make good crystal oscillators—especially where the crystal oscillates in its parallel-resonant mode. Their square-law transfer characteristic can be advantageously used to design efficient frequency-doubling stages. The classic oscillator circuits (Hartley, Colpitts, Clapp, Miller, etc.) are all as adaptable to the JFET as to tubes. Indeed, the tube-oriented amateur or designer is likely to feel at home with JFET RF applications. One difference, however, is that there is no counterpart of grid current in FET oscillators or in RF amplifiers. The junction diode input section must never be conduction biased. Driving the gate sufficiently positive in an n-channel JFET to cause gate current will limit the RF output in the drain circuit, rather than increase it. Otherwise, the JFET is very tubelike. Recall, in this regard, that many, if not most, tubes operate as depletion devices; they require a reverse bias to establish an oper-

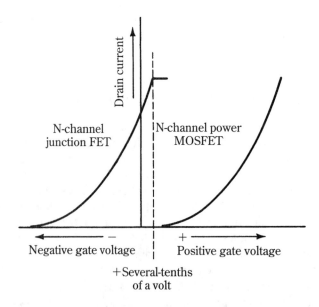

3-1 Operational modes of FETs that are used for low- and medium-power RF functions. The junction FET operates in the depletion mode. The power MOSFET operates in the enhancement mode.

ating point for class-A operation. The JFET is sometimes used in class-A mode in such applications as VFOs and buffer amplifiers. Where output power and efficiency are important, class-C type operation predominates (without gate current).

Unless power JFETs are developed for RF applications, it is likely that the now-available power MOSFET devices will displace most JFETs. In Fig. 3-1, the operational modes of JFETs and power MOSFETs are shown. Small-signal MOSFETs often operate in both the depletion and enhancement regions, but are not relevant for our purposes because of their tiny power capabilities). Notice that the JFET might be biased slightly positively. This is only on the order of several tenths of a volt, because any higher positive voltage will forward-bias the pn junction and thereby limit the operation of the device. Various tubes can be classified as depletion or enhancement devices, with most sharing both modes. Bipolar transistors operate in the enhancement region of Fig. 3-1, although this terminology is not used to describe their operational mode. The same applies to zero-bias tubes. The significance of these operating modes pertains primarily to biasing arrangements. The dynamic characteristics of two devices can be quite similar, even though their operational mode and bias circuits are different.

Amateurs have been traditionally adept at squeezing more power out of devices than is specified by manufacturer's ratings. With the JFET, however, they have not been successful. This has been frustrating because of the other fine qualities of the device. Overseas technology has developed power JFETs, but not much information is yet available with regard to their performance in RF circuitry. Figure 3-1 shows why JFETs cannot be overdriven and Fig. 3-2 shows that one also runs into difficulties if an attempt is made to raise the drain voltage. The sudden discontinuity in the curves results from avalanche breakdown in the gate-source section of the device because a portion of the drain potential adds to the reverse bias of the gate. Thus, a larger heatsink, or intermittent service (such as CW), does not increase available power output.

Simple circuitry at low-power levels

Radio-frequency circuitry using JFET devices is simpler than its bipolar transistor counterparts. Actually, the JFET is usually compatible with RF, whether it was deliberately designed for such service or not. One reason is the junctionless output section of the JFET. Another is the way in which the gate-junction is operated; in its reverse-bias mode. The net result is that a grounded-gate circuit (Fig. 3-3), can perform well over a frequency range of 20 to 200 MHz. The input impedance is actually much higher than 50 Ω and displays virtually no reactive component over this frequency range. Thus, the input network problem, so important with bipolar designs, is of little consequence here. Transformer T1 in the output circuit is a transmission-line type, which provides a 16-to-1 impedance step down. The beaded leads function as RF chokes at the higher frequencies. Among the many ferrite beads that are suitable are the Carbonyl J units that are made by the Pyroferric Company. Generally, high permeability and fairly high loss (to keep the Q low) are the criteria for selecting such ferrite beads.

A surprising feature of such a circuit as Fig. 3-3 is its respectable performance as a linear amplifier. This appears to contradict the well-known fact that the large-

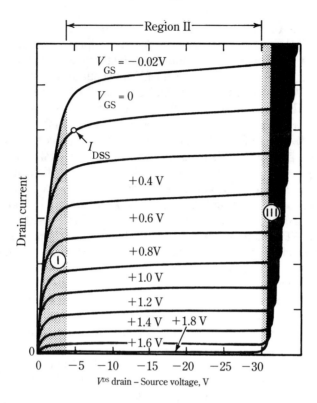

3-2 Output characteristics of a p-channel JFET showing drain-voltage limitations. Region III represents avalanche breakdown in the gate-source pn junction.

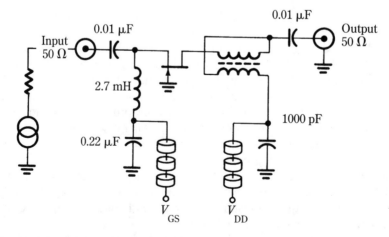

3-3 JFET building-block buffer or low-power output amplifier. Grounded-gate circuit is noteworthy for its simplicity and its broadband characteristics. _{Siliconix}

signal operation of a JFET is anything but linear. However, the nonlinearity of the device follows a nearly true square-law relationship. This does not give rise to the third-order intermodulation distortion by which linear amplifiers are evaluated. Thus, by application of an appropriate bias voltage, V_{GS}, satisfactory performance (such as depicted in Fig. 3-4) can be attained.

As already pointed out, the power ratings of most JFET devices have been quite low. A JFET, such as the Siliconix U322, is rated for a power dissipation of 3 W with its TO-5 package at 25°C. This is a relatively high-powered unit!

The reason that grounded-gate amplifier circuits are popular with the JFET designers is that stability and flat frequency response are readily forthcoming without the drawback displayed by the bipolar transistor in similar circuits. This drawback is the very low input impedance, which the forward biasing of the emitter-base junction engenders. All things considered, it is tantalizing that multiwatt JFETs are not generally available for RF purposes. Certain overseas developments suggest the possibility of such devices in the near future. However, the dramatic development of the power MOSFET now makes such a "crash program" less compelling.

Power FETs: "solid-state tubes"

A true technological breakthrough has been made with regard to MOS field-effect transistors. These devices are no longer restricted to signal-level power ratings. Although the popular consensus of field-effect transistors as flea-power devices with inherent electrical fragility might, indeed, have had a factual basis, it is also fact that we are now in an era of FETs with current ratings of approximately 30 A and voltage

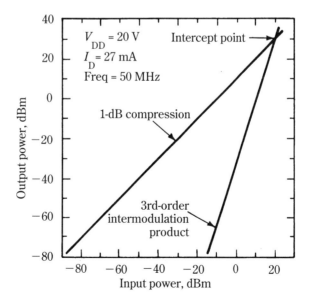

3-4 The linear amplifier performance of the grounded-gate amplifier of Fig. 3-3. The output current of the JFET relates to input voltage by a nearly true square law, but third-order intermodulation products tend to be low. Siliconix

ratings of 600 V or more. That these devices are of the insulated-gate (MOS) variety with rugged electrical characteristics is incredible. Then, to learn that such FETs can operate efficiently as power amplifiers and oscillators well into the UHF region requires a rethinking of attitudes. Not only have MOSFETs evolved into capable RF power devices, but they display the following advantages over bipolar types:

- They do not suffer from secondary breakdown, because there is no pn junction in the output section (i.e., the drain-source circuit). Also, there is no hot-spotting.
- They are not vulnerable to thermal runaway. This feature stems from the positive coefficient of channel resistance, with respect to channel current. Also, the transconductance decreases with rising temperature, which tends to shut the device down.
- There is no current hogging; units can be directly connected in parallel without ballast resistances.
- Frequency capability is ultimately limited by stray reactance, rather than by charge-storage effects. This makes possible a very high gain-bandwidth without the greatly detrimental trade-offs of voltage or power ratings.
- The input impedance can be said to be tube-like. This simplifies certain input tank problems and is less demanding on the drive circuitry.
- Broadbanding is often more readily accomplished.
- Low-frequency instability is less likely because operation generally occurs at or near the maximally available power gain of the device.
- The VSWR of the load does not endanger the device.
- Excellent proportionality exists between input voltage and output current. Thus, the device is inherently applicable to linear RF power amplifiers. It is not always easy to obtain such linearity from bipolar RF power transistors.
- More reasonable impedance transformation is often possible because of operation at higher voltages and lower currents.
- Cost is at least comparable to ordinary power transistors and might be lower than true RF types of bipolar transistors.
- Biasing tends to be less critical.
- No varactor action is in the output (drain-source) circuit. This reduces harmonic generation.
- Ultrafast switching performance makes switching-type RF amplifiers feasible, in which theoretical efficiency is 100% (maximum theoretical efficiency of the ideal class-C amplifier is about 85%).

Other unique features of the VMOS power FET A curious, but potentially useful characteristic of the VMOS power FET is its inordinately low noise figure (near 2.5 dB at 150 MHz). To be sure, the noise figure does not have much importance for the designer of power output stages. However, the low noise figure of this device in conjunction with yet another strange feature suggests novel applications. This feature is its substantially constant input and output impedances from several microwatts to several tens of watts or more of output power. This, of course, bodes well for linearity in single-sideband amplifiers and for frequency stability in oscillators, but these features also suggest an application that would be almost unthinkable with bipolar RF transistors.

The great emphasis on transceivers during the past two decades has taxed the circuit designer's ingenuity for techniques that lead to simplicity and cost effectiveness. From the preceding paragraph, it is apparent that the very same VMOS FET RF final-amplifier stage in a transceiver could also function as a receiver RF input amplifier. This, of course, requires a reorientation of attitude; you certainly would not be conditioned to think of a bipolar power transistor in such a dual role.

It appears, also, that the VMOS power FET should be ideally suited for crystal oscillator circuits. The low input impedance of bipolar transistors has always been an adverse factor in such applications. And, other things being equal, the VMOS power FET should also be a better candidate for attaining frequency stability in VFOs. This device also appears to be better suited for achieving isolation in buffer amplifier stages. Additionally, its higher input impedance is less likely to exert a loading effect upon the VFO or the preceding stage. Many VFO designs have used small-signal junction FETs to accomplish these various objectives. The low power-handling capability of these JFETs has often posed obstacles, however.

Yet another unique feature of VMOS power FETs should be of interest to the far-sighted designer or imaginative experimenter. The input of these devices can directly interface with CMOS, TTL, DTL, and MOS logic families. Thus, digital circuitry could be used to accomplish switching operations in class-D fashion. Under such conditions, an appropriate LC tank, or harmonic filter in the output circuit, would extract the fundamental frequency from the square wave. Because the risetimes and falltimes can be on the order of 4 nanoseconds, (ns), such an approach merits consideration.

It is anticipated that as the technical community outgrows its obsolete notion that MOS technology is limited to flea-power devices, these and other unique applications will naturally evolve.

V-groove structure of the power FETs

The difference in fabrication between the early power FET and the conventional flea-power MOSFET is illustrated in Fig. 3-5. The V-groove gate structure (Fig. 3-5A) allows the controlled source-drain current to follow a vertical path through the four semiconductor regions. Notice that the drain region occupies a large area. This enables the case-to-heatsink thermal resistance to be low. Because the V configuration involves two channels, the single metallized gate controls two vertical current paths. The n-region is an epitaxial layer and behaves as a space charge region, which enables much higher drain voltages to be applied than in the case of the ordinary MOSFET structure (Fig 3-5B). Another beneficial effect of the n-layer is a considerable reduction in output capacitance and in drain-gate feedback capacitance—of prime significance in RF circuits.

The power MOSFET operates in the enhancement mode; that is, output current is cut off when the gate has the same potential as the source. A positive gate bias on the order of 5 to 15 V is required for rated channel current in the source-drain cir-

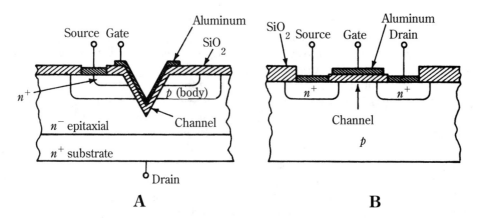

3-5 Comparison of the power MOSFET structure with ordinary MOSFET geometry. A. The VMOS structure of the new power units. Source-drain current path is vertical through the four semiconductor regions. B. A previous MOSFET configuration—limited to signal-level power.

cuit. However, unlike either tubes or bipolar power transistors, this positive bias is not accompanied by dc bias current. Therefore, the dc input impedance is very high, comparable to that of tubes that operate with negative grid bias. A corollary of this is that the power MOSFET requires very little power from its driver stage. No input current is similar to grid current in a class-C tube amplifier.

Those familiar with low-power insulated-gate transistors are aware of their vulnerability to destruction from static electricity. Merely handling one of these devices is often enough to blow out the gate structure.

Some of the power MOSFETs have internal monolithically fabricated protection diodes that are associated with the gate structure. Those that do not require very careful handling. However, once in their circuits, the susceptibility to gate damage from transients or static charges is greatly reduced. Nonetheless, it is unwise to use an ungrounded electric soldering iron around a circuit that uses the power MOSFET. Power MOSFETs that are specifically intended for RF applications generally do not incorporate internal gate protection in order to avoid the adverse effects that protective diodes have on the input impedance.

Some revealing features of the power MOSFET

Suppose that, as a seasoned veteran of bipolar transistor circuits, you encounter a set of operational curves, such as those shown in Fig. 3-6. Casual inspection might not stimulate unusual interest. However, take more than a superficial glance at these curves. The output characteristics depicted in Fig. 3-6A appear familiar enough; they resemble those of pentode tubes and of bipolar transistors. You rarely see such curves for bipolar transistors representing the most often used common-emitter connection. Rather, they resemble those generated in the common-base circuit. Common emitter curves are less evenly spaced and have appreciable slopes.

Unfortunately, common-base circuitry often presents awkward impedance levels—especially in RF work. The flat curves of Fig. 3-6A are for the "common-emitter"

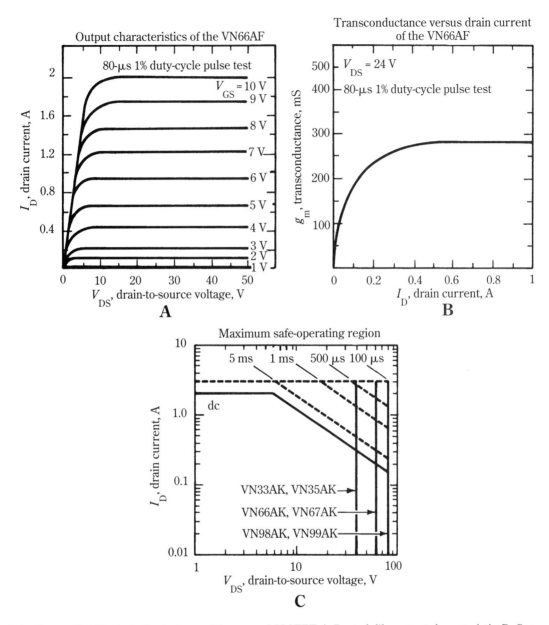

3-6 Curves that illustrate the features of the power MOSFET. A. Pentodelike output characteristic. B. Constant transconductance over appreciable operating range. C. No secondary breakdown limitation in SOA curves. Siliconix Corp.

(now called "common-drain") connection of the power MOSFET. The horizontal aspect of these curves indicates very-high output resistance. The even spacing of many of them indicates an extensive linear region of operation. You might rightfully infer that the power MOSFET is a good candidate for a class-AB linear RF power amplifier.

Figure 3-6B reveals similar information in a different way. Here, you can see that the transconductance of the device is flat over an appreciable operating range. Notice that the flat region embraces the evenly spaced operating range of Fig. 3-6A. Also, notice that the transconductance is in the vicinity of 275 mS, a "hot" tube indeed! It is true that power bipolar transistors exhibit tens of siemens of transconductance. However, the power gain or current gain of bipolar transistors is low or moderate because of their current-consuming input circuits. Another way of making such comparisons (not revealed by these curves) is to contemplate the equivalent beta of power MOSFETs, which is on the order of 1 billion! In this respect, the power MOSFET can be likened to a "solid-state tube."

In the safe-operating-area graph of Fig. 3-6C, something is conspicuous by its absence, a boundary imposed by secondary breakdown. As already mentioned, the power MOSFET has no such limitation because there are no pn junctions in its output section or channel. Thus, the SOA curves are not unlike those that could be drawn for a simple passive element, such as a resistor. A little thought reveals that even a resistor has a maximum safe current, a maximum safe voltage, and an operating region bounded by power dissipation considerations. The absence of secondary breakdown or of hot-spotting regions is one of the salient features of the power MOSFET. The practical manifestation is that the device is not likely to suffer catastrophic destruction because of an unresonated tank circuit or because of high VSWR presented by an antenna feedline. By the same token, it will not be vulnerable to overmodulation or transients.

Example of MOS power capability   Those accustomed to the milliwatt range of ratings of small-signal MOSFETs, or the fractional-watt operating levels of JFETs, will be surprised by the specifications of the VN84GA power MOSFET (Fig. 3-7). Although this particular power MOSFET is not optimally designed or packaged for VHF applications, its 50-ns switching time suggests good performance in the popular amateur high-frequency bands. Another favorable factor portending satisfactory RF operation is the absence of an input Zener diode. Notice the forward transconductance of 2 S; compare this with the "hottest" transmitting tube you can find. It will be evident that the oft-used adjective "tubelike" might be an injustice.

Input circuit of the VMOS power FET

The VMOS power FET shares with small-signal MOS transistors vulnerability to destruction from static charges applied to its gate. Once the device is in its circuit, this danger might no longer exist, but during initial handling, consideration must be directed to static-generating conditions, such as walking on synthetic fiber carpets, wearing crepe-sole shoes, etc. Also, it is wise to watch out for leakage currents from soldering irons during installation. Many VMOS power FETs have monolithically built-in Zener diodes in the gate-source circuit and are relatively immune to damage from static charges and leaky power line transients. Unfortunately, this generally applies only to those devices that are intended for audio, servo, and low-frequency service. Although the protection conferred by the input Zener diode would be welcome in RF applications, the side effects pose difficulties. These include increased power consumption from the driver stage and increased input capacitance. Also, the Zener diode, itself, is vulnerable to burnout.

Junction field-effect transistor (JFET)

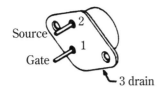

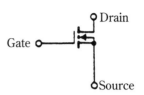

Absolute maximum ratings

Maximum drain-source voltage ....80 V
Maximum drain-gate voltage ....80 V
Maximum continuous drain current ....12.5 A
Maximum pulse drain current ....15 A
Maximum dissipation at 25° C case temperature ....80 W
Linear derating factor ....1.56°C/W
Temperature (operating and storage) .... −55 to +150° C
Lead temperature
   (1/16" from case for 10 seconds) ....300°C

ELECTRICAL CHARACTERISTICS (25°C unless otherwise noted)

		Characteristic	Min	Typ	Max	Unit	Test Conditions
1	S T A T I C	BV_{DSS} — Drain-Source Breakdown	80			V	V_{GS} = 0V, I_D = 100 mA
2		$V_{GS(th)}$ — Gate Threshold Voltage		2.5			V_{DS} = V_{GS}, I_D = 10 mA
3		I_{GSS} — Gate-Body Leakage			100	nA	V_{GS} = 10 V, V_{DS} = 0
4		I_{DSS} — Zero Gate Voltage Drain Current			1.0	mA	V_{DS} = Max. Rating, V_{GS} = 0
5		$I_{D(on)}$ — ON-State Drain Current	10			A	V_{DS} = 25 V, V_{GS} = 10 V (Note 1)
6		$R_{DS(on)}$ — Drain-Source ON Resistance		0.3	0.4	Ω	V_{GS} = 10 V, I_D = 10 A (Note 1)
7		g_{fs} — Forward Transconductance	1.5	2.0		℧	V_{DS} = 20 V, I_D = 5 A (Note 1)
8	D Y N A M I C	C_{iss} — Input Capacitance		640		pF	V_{GS} = 0, V_{DS} = 25 V, f = 1.0 MHz
9		C_{rss} — Reverse Transfer Capacitance		50			
10		C_{oss} — Output Capacitance		300			
11		t_{on} — Turn-ON Time		50		ns	V_{DS} = 20 V, I_D = 10 A (Note 2) R_L = 2Ω
12		t_{off} — Turn-OFF Time		50			

NOTES: 1. Pulse Test — 300 μs, 1% duty cycle
         2. See switching time test circuit

VNG

3-7 The operating specifications for the VN84GA power MOSFET. The data indicates a respectable power-handling capability. <small>Siliconix Corp.</small>

Although the gate exhibits essentially infinite impedance to a dc-controlling signal, this is not the case in RF applications, where input capacitance, together with the Miller effect from internal drain-gate feedback, makes the gate a relatively low impedance. However, the impedance at the gate is, nonetheless, at least an order of magnitude greater than would be encountered at the base of a bipolar transistor of similar power capability. This has much practical significance, for many of the impedance-matching problems that are associated with the inordinately low input impedance of bipolar transistors are averted. At VHF frequencies, the VMOS power FET might typically display an input impedance in the several tens of ohms, contrasted to several tenths of an ohm in bipolar power transistors.

In most RF power applications, the VMOS FET is used in the common-source circuit, which is similar to common-emitter and common-cathode configurations in bipolar transistors and tubes, respectively. Thus, the drive signal is applied to the gate. This results in stable operation from low-frequency RF through UHF frequencies. The tendency toward low-frequency oscillation experienced in bipolar transistor amplifiers is practically nonexistent with these devices.

Unlike certain other MOS devices, the VMOS power FET operates in the enhancement mode; output (drain) current is zero with no input signal. Thus, a measure of fail-safe performance is inherent in its very nature. The enhancement mode usually results in simplified bias circuitry and in a more direct approach to class-B and class-C operation. At the present writing, most development has centered on n-type devices (the counterpart of npn bipolars). Near-optimum class-C performance generally obtains from zero-bias operation (i.e., with the gate at source potential). From the standpoint of circuit simplicity, this is a desirable feature.

Typical input impedance versus frequency for a power MOSFET is shown in Fig. 3-8. This is for a series-equivalent circuit; as with a conventional capacitor, the impedance can be represented as either a very high resistance in parallel with a capacitance (parallel-equivalent circuit) or a very low resistance in series with a capacitance. The reactive component will be seen to have a constant slope; that is, it is a straight line. This has a special significance on a log-log plot; it implies that the capacitance does not change with frequency. This contrasts to the frequency-dependent input capacitance of bipolar transistors. A related factor, not indicated on this graph, is that the input capacitance of the power MOSFET does not change with drive level, again contrasted with the situation in bipolar transistors. If you were not otherwise informed, the reactance curve of Fig. 3-8 could be construed to represent a high-quality capacitor. Indeed, the combined impedance plots, reactance and resistance, could be so interpreted.

When first encountered, the common-gate input impedance of the power MOSFET often evokes surprise. As seen in Fig. 3-9, the input impedance can be purely resistive out to several tens of megahertz. This follows directly from the common-source curves of Fig. 3-8, but the relationship requires the mathematical, rather than the intuitional approach. The plots of Fig. 3-9 also show why common-gate performance at higher frequencies is often plagued by instability, for both the resistive and reactive components of input impedance then depart from their near-ideal levels. Such departure is likely to cause an impedance mismatch in such an amplifier (at best) or high-frequency oscillation (at worst).

Junction field-effect transistor (JFET)   87

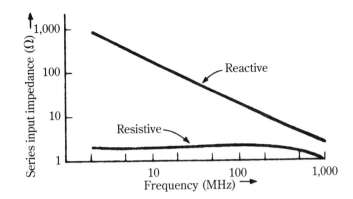

3-8 The common-source input impedance of the typical RF power MOSFET. Curves represent a nearly "pure" capacitance in series with a low and essentially constant resistance.

Output circuit of the VMOS power FET

From a practical standpoint, the salient feature of the output (drain-source) circuit of the VMOS power FET is that there are no pn junctions. Rather, there is only a relatively simple conductive path. When compared to the bipolar transistor, the implications of this junctionless output circuit are significant. First, the destructive phenomenon of secondary breakdown is absent. Also, because of the lack of pn junctions, there is no varactor effect from voltage-dependent pn capacitance. This results in an easier-to-filter harmonic spectrum in class-B and class-C amplifiers. The output circuit is bilateral in its conductive properties so that no protective diode needs to be used in class-D or class-F amplifiers. This is important, for such diodes would degrade the efficiency of these high-frequency switching amplifiers. This is one of the reasons why the bipolar transistor has not been successfully used in these otherwise efficient operational modes for RF amplifiers.

Another closely related aspect of such a junctionless output circuit is the absence of thermal runaway. Whereas in a bipolar transistor, excessive junction temperature regeneratively increases output current, and thereby increases temperature further,

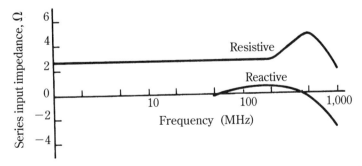

3-9 The common-gate input impedance of the typical RF power MOSFET. The flat resistive input impedance portends readily attained wideband performance up to several tens of megahertz.

the very opposite situation prevails in the VMOS power FET. Here excessive temperature, from whatever cause, reduces the conductivity of the output circuit, which thereby tends to self-correct the overheated operation. In any event, the device does not drive itself to thermal destruction. A corollary of this desirable characteristic is that VMOS FETs can be paralleled without fear of current hogging by one of the units. No ballast resistances are needed in the parallel combination, for the internal mechanism that prevents thermal runaway also mitigates against current hogging. This property is summed up by stating that the device has a negative coefficient of output current, with respect to temperature. In the technical literature, you sometimes find allusion to the "positive temperature coefficient" as being responsible for absence of thermal runaway. What is meant here is "a positive coefficient of *output resistance*," with respect to temperature.

It should not be construed that the absence of thermal runaway stems from the conductive properties of the drain-source silicon element itself. Indeed, silicon increases its conductivity with temperature. What happens in the overall device, however, is that the transconductance decreases with temperature. This counteracts any tendency of output current to runaway when the FET becomes hot.

The transfer characteristics of the power MOSFET are best described by transconductance, the same parameter that is so useful with tubes. Transfer curves, such as shown in Fig. 3-10, provide much information for the designer. Transconductance is readily obtained as the ratio of change in drain current, caused by a small change in gate voltage, while drain voltage is held constant. Ap-

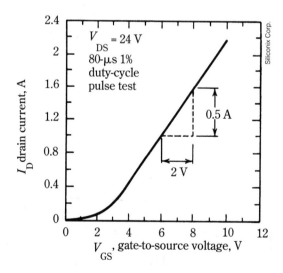

3-10 The transfer characteristic of the VMP-4 power MOSFET. This curve depicts a threshold of approximately 2 V and very good linearity for drain currents that exceed 400 μA. The transconductance is:

$$\frac{0.5 \text{ A}}{2.0 \text{ V}} = 250{,}000 \ \mu s$$

plying this concept to Fig. 3-10, the transconductance in the linear region is readily found to be 250,000 μS.

Class-D RF amplifiers

The operating modes that have been most common with tube amplifiers have been class B (or AB) and class C. The former is used to provide linear amplification of SSB signals and the latter is useful for CW, AM, and FM modes. The same has been true of transistor amplifiers, except that class-A operation is often used at low power levels. With both tubes and transistors, class-B amplifiers are sometimes used for CW and FM, despite its somewhat lower efficiency, compared to class-C operation. Also, the class-B (or AB) amplifier, because of its linearity, can provide power boost for AM. With both tubes and transistors, the class-C operating mode has enjoyed universal application in frequency-multiplier stages.

The maximum theoretical efficiency for class-A stages is 50%, that for class-B operation is 78.5%, and class-C amplifiers can attain efficiency percentages around 85%. In all instances, the cited efficiencies prevail only when the amplifiers are driven to maximum output; at lower outputs, the efficiency is lower.

But, there is yet another mode of amplifier operation, which is capable of even greater efficiencies than is ordinarily possible with the class-C amplifier. This operating mode, class D, causes the amplifying device to behave as a near-ideal switch. Such a switch is on half of the time and off the remaining half. To perform this way, the transition time between switching states must approach zero. The salient feature of an ideal switch is that there is no power loss when it is on because the on resistance is construed to be zero. It has no power loss when it is off because its off resistance is construed to be infinite. If the switching transition time is zero, no power loss can occur between the states. Of course, the output of such a switch is a square wave and it must be adequately filtered so that the load is presented with a clean sine wave. Such filtering can be best accomplished with the aid of a harmonic filter inserted between amplifier and load.

Although it is commonly said that the theoretical maximum efficiency of the class-D amplifier is 100%, this statement is not quite applicable to RF amplifiers because the load only uses the first harmonic (fundamental frequency) of the square wave. The higher harmonics are prevented from reaching the load. So, the utilization factor is not 100%. However, the unused energy is not dissipated as heat in the switching device, and the dc power supply is not called on to supply such dissipation. The fact remains that the class-D amplifier is the most efficient type of all. Figure 3-11 shows the basic class-D RF power amplifier.

The question naturally arises as to why the class D mode was virtually never used with electron tubes and has been rarely encountered with bipolar transistors. In the case of tube amplifiers, the departure from an ideal switch robs much of the incentive to operate in the class-D mode. This is because of the appreciable plate-cathode voltage drop—even when the tube is hard driven. Also, the impedance levels in tube circuits make it difficult to develop and preserve good square waves at high power levels. Hams have sometimes inadvertently approached class-D operation by overdriving their class-B and class-C amplifiers. Unfortunately, the real benefits of this operational mode accrue only when you make deliberate efforts to produce very rapid switching transitions.

Bipolar transistors simply have not been sufficiently fast-switching elements to

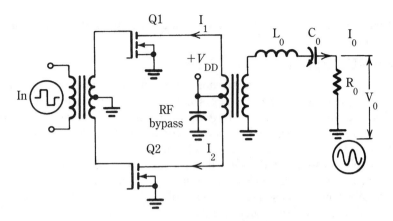

3-11 The basic arrangement of class-D RF power amplifier. High efficiency is achieved because the power FETs are either on or off, never in an in-between (dissipative) state.

produce suitable square waves at repetition rates of 2 MHz or greater. The very efficient performance obtained from nonresonant switching-type power supplies generally uses switching rates on the order of 20 to 40 kHz.

The advent of the power FET has made class-D operation feasible for frequencies at least up to 30 MHz. These devices, unlike bipolar transistors, are not plagued with minority charge-storage problems. Although the drain-source voltage drop exceeds the collector-emitter drop of good RF bipolar transistors, it is still quite low. By using as high a dc voltage as permissible, the effect of the drain-source voltage drop is of minor consequence. It is the ability of the power FET to produce nearly instantaneous risetimes and falltimes that renders it a superb device for class-D RF service.

Actually, two ways cause the power FET to develop a square wave in its drain circuit. The most obvious has already been alluded to: a square wave of voltage is applied to the gate, and the drain-source circuit simply follows suit in switching from the on to the off state of conduction. Another technique is also simple, but not quite so obvious. It involves inserting a quarter-wave transmission-line section in the drain circuit. The gate need not be impressed with a square wave. Rather, half-sine waves can be applied to the gate in much the same manner used in class-B operation. Figure 3-12 illustrates the basic circuitry. Although the power FET is caused to function in a similar manner to the class-D amplifier, such a circuit is said to operate in the *class-F mode*. By referring to the waveforms of Fig. 3-13, the operation of the class-F amplifier can be explained.

Assume you have a sine-wave input. If it was not for the quarter-wave line, the drain voltage would be sinusoidal and the FET would operate in class B. In class-B operation, efficiency is limited by, among other things, that both drain voltage and drain current exist at times. During such times, there is dissipation in the drain-source circuit of the FET. If, however, drain voltage could be converted into a square wave, drain voltage and drain current would never coexist. Then, their product would be zero throughout the RF cycle, and there would be no power dissipation in the drain-source circuit of the FET. Notice that essentially the same reasoning applies here as in the class-D amplifier of Fig. 3-11. It is evident that the quarter-wave

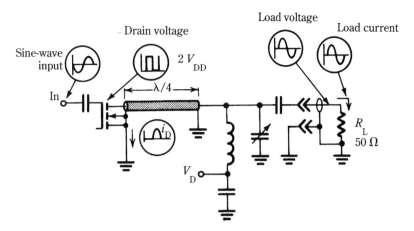

3-12 The basic arrangement of class-F RF power amplifier. The ¼-λ transmission line causes the drain voltage to be a square, rather than a sine wave. Because of this, the FET operates in similar fashion to the class-D circuit of Fig. 3-11.

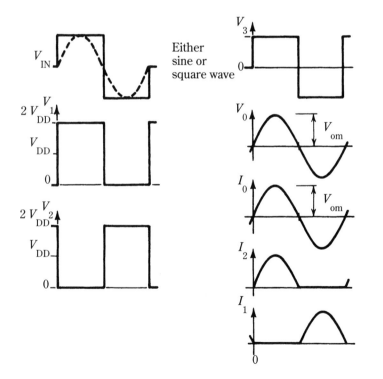

3-13 A waveform diagram of the class-D RF power amplifier. Notice that the drain voltage of a particular FET is zero while drain current flows. Such an idealized situation implies zero drain-circuit dissipation.

line, although neither an active nor a nonlinear element, exhibits wave-shaping properties.

A quarter-wave line inverts the impedance of its load at odd harmonics of the fundamental frequency. For example, if a quarter-wave line is loaded by a short circuit, the generator end of the line will "see" an open circuit (an infinite impedance) at f, $3f$, $5f$, $7f$, $9f$, and so on. Notice that the fundamental frequency (f) is, itself, an odd harmonic. One widely used manifestation of this transmission-line behavior is the use of quarter-wave lines to simulate parallel-resonant circuits in order to provide high impedance at the fundamental frequency.

Because the quarter-wave line is used in conjunction with a simple parallel-tuned tank, such a parallel-resonant circuit behaves as a high impedance to the fundamental frequency, but as a short circuit to all other frequencies. That is, it supports frequency f, but it rejects all its harmonics. Now focus your attention on the drain of the FET and see what happens as a consequence of the combined actions of the quarter-wave line and the parallel-resonant LC circuit.

All odd-harmonics are supported at the drain—even though they experience short-circuits at the LC tank. This stems from the impedance-inverting property of the quarter-wave line. For f, itself, the drain "sees" the input impedance of the line as

$$\frac{(R_o)^2}{R_L}$$

where R_o is the line's characteristic impedance. Although f is thereby supported at the drain, its amplitude is not as great as it could be if the input impedance of the line was as high as it is for all of the odd harmonics of f. This causes the concave depression in the top of the drain-voltage waveform. It, however, is of little practical significance; essentially, class-D operation is simulated with the added benefit that sine-wave drive can be used.

Thus, the drain circuit is caused by the line and LC tank to support f and its odd harmonics. Even at harmonics of f, the line simply reproduces the short-circuit that is imparted by the LC tank. In this way, the requisite square wave is synthesized at the drain. The synthesis stems from the appropriate combination of the Fourier constituents ("building blocks") of a square wave: a fundamental frequency and a large number of its odd harmonics. The mathematical justification of this phenomenon also requires favorable conditions of magnitude and the phase of the harmonics. At frequencies where a linear section of quarter-wave line is not physically practical, coiled-up miniature coaxial cable can be used. In any event, the line must electrically be a quarter wavelength.

A family of MOSFET giants

The IRF35X family of power MOSFETS made by International Rectifier Corporation are worthy of consideration, not merely because of their inordinately high power dissipation ratings (150 W), but because of their simultaneously high dc operating voltages. The 350- and 400-V drain-source voltage rating carried by these devices should greatly relieve the problems that accompany low-impedance output networks, the scourge of high-power amplifiers that use bipolar transistors. Because the required impedance of an output network is nearly proportional to the square of the dc operating voltage, these MOSFETS can be implemented with output networks that have impedance levels on the order of 250 times greater than that for a bipolar transistor of similar power capability, but operating from a nominal 25-Vdc source. Except for lack of the familiar filament glow, the nature of such a MOSFET RF amplifier is, indeed, suggestive of tube circuitry.

Admittedly, these power MOSFETS are neither characterized nor intended for RF applications. It is generally wiser to use RF power devices that have been optimized for specific performance parameters. However, experimentation is often the precursor of devices and techniques with specific engineering orientations. These MOSFETS should whet the appetite of the experimentally inclined.

The electrical characteristics of this family of power MOSFETS are given in Fig. 3-14. Curves that depict the salient operating characteristics are shown in Fig. 3-15. You should not be discouraged by the turn-on and turn-off delay times. These apply to switching circuits, where time must be allotted to charge and discharge the gate input capacitance. In RF applications, the gate capacitance will be incorporated as part of the input network and there will be no such "delay."

It appears likely that these interesting devices could readily be used in narrow-band class-C amplifiers for service in the high-frequency amateur bands (2 to 30

Electrical Characteristics @ T_C = 25°C (Unless Otherwise Specified)

	Parameter	Type	Min.	Typ.	Max.	Units	Conditions
BV_{DSS}	Drain – Source Breakdown Voltage	IRF350 IRF352	400			V	V_{GS} = 0
		IRF351 IRF353	350			V	I_D = 1.0 mA
$V_{GS(th)}$	Gate Threshold Voltage	ALL	1		3	V	$V_{DS} = V_{GS}$, I_D = 1 mA
I_{GSS}	Gate – Body Leakage	ALL			100	nA	V_{GS} = 20V
I_{DSS}	Zero Gate Voltage Drain Current	ALL		0.1	1.0	mA	V_{DS} = Max. Rating, V_{GS} = 0
				0.2	4.0	mA	V_{DS} = Max. Rating, V_{GS} = 0, T_J = 125°C
I_D (on)	On-State Drain Current	IRF350 IRF351	11			A	V_{DS} = 25V, V_{GS} = 10V
		IRF352 IRF353	10			A	
R_{DS} (on)	Static Drain-Source On State Resistance	IRF350 IRF351		0.25	0.3	Ω	V_{GS} = 10V, I_D = 5.5A
		IRF352 IRF353		0.3	0.4	Ω	
g_{fs}	Forward Transconductance	ALL	5.0	9.0		S (℧)	V_{DS} = 100V, I_D = 5.5A
C_{iss}	Input Capacitance	ALL		3000	4000	pF	V_{GS} = 0, V_{DS} = 25V, f = 1.0 MHz
C_{oss}	Output Capacitance	ALL		400	600	pF	
C_{rss}	Reverse Transfer Capacitance	ALL		100	200	pF	
t_d (on)	Turn-On Delay Time	ALL		40	60	ns	I_D = 5.5A, E_1 = 0.5 BV_{DSS}
t_r	Rise Time	ALL		100	150	ns	T_J = 125°C (MOSFET Switching times are essentially independent of operating temperature.)
t_d (off)	Turn-Off Delay Time	ALL		300	400	ns	
t_f	Fall Time	ALL		100	150	ns	

Thermal Characteristics

$R_{\theta JC}$	Maximum Thermal Resistance Junction-to-Case	ALL		0.83		deg C/W	

3-14 The electrical characteristics of a family of high-power MOSFET devices. Although not intended for RF applications, the experimentally inclined will find these devices worthy of consideration. International Rectifier Corp.

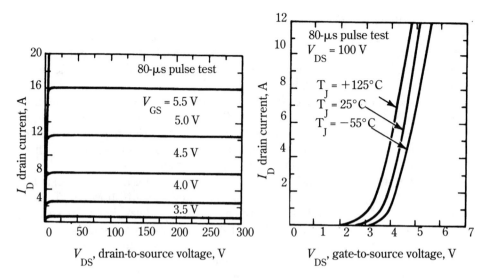

3-15 Basic characteristics of giant power MOSFETs. Evidenced in these curves are the pentodelike operating mode, the exceedingly high transconductance, and the possibility of designing an amplifier circuit with a high-impedance output network (because of the relatively high-voltage operating range compared to bipolar transistors.) *International Rectifier Corp.*

MHz). Probably, neutralization would be required. Also, amateurs have found MOSFETS capable of performing well in linear amplifiers for SSB transmitters.

The symbology and nomenclature referring to a "junction" should not be taken to imply a pn region (for example, T_J, $R_{\Theta JC}$, and actual use of the word "junction" in Fig. 3-14). This is a carry-over from the terminology used in specifying bipolar transistors. For practical purposes, you can think of internal operating temperature, rather than the alleged "junction" temperature. Another interpretation is to think of the dielectric "junction" between the gate and the drain-source structure.

Three circuit configurations

Bipolar transistors are useful for RF power applications in the three circuit configurations: common emitter, common base, and common collector. Correspondingly, tube RF circuits can use the common-cathode, common (grounded) grid, and common-plate connections. At the present writing, the power MOSFET is most frequently used in RF circuits that are configured around the common-source connection. A salient feature of common-source operation is that amplifier stability improves as the frequency goes higher into the VHF and UHF region.

Do not assume that other connection modes are ruled out for power MOSFETs. For example, the relative fragility of the gate structure is no impediment to common gate operation if you take care to maintain voltage and current ratings. Although input impedance will then be low, it will be much higher than for the similar operation (common base) and ratings of a bipolar transistor. However, the stability of common-gate amplifiers is much better in the low- and high-frequency spectrum than at VHF and UHF frequencies. Thus, this situation is the opposite of that which prevails with the practical implementation of bipolar transistors. Common-gate amplifiers are noteworthy in that they can exhibit constant and essentially resistive input impedances over many frequency octaves.

The experimentally inclined should also investigate the attributes of the common-drain circuit. Although this configuration has been largely played down or ignored, a similar situation long endured for RF applications of bipolar power transistors. At least one major semiconductor firm now produces a wide line of bipolar power transistors that are specifically intended for the common-collector configuration at microwave frequencies. Excellent performance can be obtained, despite much literature that assigns inferior status to this circuit mode for RF work. Broadband (2- to 50-MHz) amplifiers that are designed around push-pull common-drain circuits have worked out well at the several-hundred watt level. Low-voltage MOSFETs should be used in order to avoid gate puncture. Good linearity can be obtained for SSB operation.

Disadvantages of the power MOSFET

Practical experience and basic principles both teach that desirable features in devices are often traded off for some less-than-favorable characteristics. Surely, the impressive behavior of the power MOSFET must be accompanied by some disadvantages, when compared with the bipolar transistor. Indeed, the following performance shortcomings can be ferreted out:

- The "saturation voltage" is higher than is generally attainable in equivalent RF power bipolars. Thus, saturation voltages in the vicinity of 3 V are not uncommon. This, of course, reduces the effective use of the dc supply and increases device dissipation.
- Input capacitance is relatively high. It is more "visible" than in a bipolar transistor, where the input capacitance is somewhat swamped out by the low resistive component of the input impedance.
- The breakdown mechanism of the gate is damaging. Thus, overdrive (overbiasing) can puncture the thin oxide-dielectric film of the gate. This contrasts to the junction FET, where gate breakdown is not necessarily injurious. Vulnerability to gate puncture increases with drain voltage.

In addition to gate damage caused by inappropriate operating conditions, the gate is subject to destruction from static electricity during handling. Once successfully installed in its circuit, the device is relatively immune from the effects of static charges. Take care to prevent leakage currents from soldering irons from passing into the gate circuit; vulnerability to damage from this often unsuspected source is much greater than with bipolar transistors.

Although the diode protection of the gate circuit reduces danger at low frequencies, by the time you get into the tens of megahertz region, the side effects and performance degradation caused by protective diodes render them less desirable than taking a little extra care in handling and operation.

Avant-garde FETs and RF techniques

The continuing evolution of power-FET technology is bound to affect high-frequency techniques. Some of the circuit applications of RF power amplifiers that use these devices did not exist just a few years ago; you could easily indulge in speculations of the science-fiction kind in prognostications of the near future. Because the theme of this book primarily deals with the practical and the realizable, the space will be devoted to neither mere extrapolation of the present state of the art, nor to "blue-sky" achievements. However, the following FET devices and techniques are representative of commercially available devices, which have rather recently made their debut. They will be of interest to those who are motivated to keep pace with ongoing developments and to the experimentally inclined:

- A number of manufacturers have introduced lines of p-channel power MOSFETs. These devices are similar to pnp bipolar transistors insofar as concerns the polarity of operating voltages. Surprisingly, these newer devices often have comparable power ratings to established n-channel devices. Even better, they have, in many instances, been designed to have very similar parameters to specified n-channel devices. The availability of such paired devices makes possible simple implementations to push-pull amplifier circuitry. An example of such a complementary-symmetry amplifier is shown in Fig. 3-16. Although FET connections are more suggestive of a parallel than a push-pull arrangement, actual operation occurs in true push-pull fashion. The great appeal of such circuitry will probably be with the designer of ultrasonic equipment. The simple drive requirement and the single-ended output tank configuration also make this application worthy of consideration as a transmitter output amplifier.

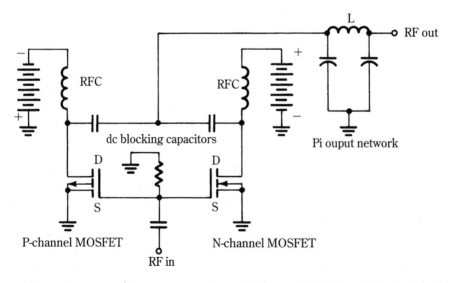

3-16 Complementary symmetry push-pull amplifier using MOSFETs. Although a dual dc supply is needed, this scheme provides benefits of push-pull operation with the simplicity of single-ended drive and output circuitry.

- At least the following firms are now marketing p-channel power MOSFET devices: International Rectifier, Siliconix, Supertex, Intersil, ITT, and Westinghouse.
- Overseas companies have developed power MOSFETs that can be driven to full output directly from 5-V TTL logic signals. At the same time, some of these devices have 1000-V drain-source ratings. This suggests relaxations in the design of output networks. More specifically, it appears that final amplifiers that use these high-voltage FETs could be associated with pi and pi-L networks that are designed for tube circuits. This could be very advantageous in the selection and availability of practically sized inductors and tuning capacitors. Also, it should be relatively easy to obtain class-D operation from the square-wave drive signals of TTL logic. Domestic manufacturers have also placed such devices on the market.
- Another overseas development is a JFET capable of delivering 1 kW of ultrasonic power into an appropriate load. It stands to reason that such a device can also be used for low-frequency, and probably medium frequency, RF power applications. You would expect the frequency capability to at least cover the AM broadcast band.
- For those whose mission it is to extend RF power performance into ever higher regions of the microwave spectrum, the gallium-arsenide FET is the logical candidate. The frequency capabilities of solid-state power devices are indicated in a general way in Fig. 3-17.

It is interesting that FET devices are not only extending RF power performance, but are obviously motivating the designers of bipolar power devices to be more innovative than ever before. A certain degree of complacency had set in once bipolar transistors proved their competitive status with certain tubes. Now, because of the unexpected power-frequency capabilities of FET structures, it is clear that bipolar devices must excel in meaningful ways, or atrophy from lack of demand. Yet, it is quite evident that the producers of bipolar transistors do not feel that this venerable device is vulnerable to the technical obsolescence claimed in FET advertisements. What we shall see immediately ahead is enhanced competition between these two solid-state

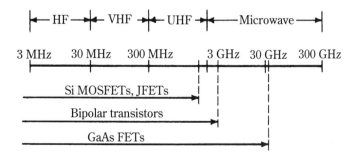

3-17 The general frequency capabilities of solid-state power devices. Notice that the gallium-arsenide FET excels the other devices in the 10- to 30-GHz region of the microwave spectrum.

power devices, as well as accelerated improvements in both. Enough time has passed so that it is fairly safe to say there will not be a "winner" per se. Rather, each will dominate in various services and applications, where unique voltage, current, frequency, cost, and other factors influence designers and hobbyists.

The IGBT: superb performance at the low end of the RF spectrum

Low-frequency (10 to 50 kHz) RF techniques are involved in important applications, including communications, navigation, induction heating, ultrasonics, and switching power supplies. For a long time, four power devices were available for generating and processing RF in this frequency range. These were the vacuum tube, the bipolar transistor, the thyristor, and the power MOSFET. More recently, the IGBT (insulated-gate bipolar transistor) has become available for such applications. It can provide a compelling blend of desirable circuit, operating, and cost-features that are not readily attainable in the other power devices.

The IGBT is often advertised as a hybrid device with the input characteristics of a power MOSFET and the output characteristics of a bipolar power transistor. Although such a description falls somewhat short of pedagogic accuracy, it is, for most practical purposes, essentially valid. The input of the IGBT appears as a capacitor that is very similar to the input of the power MOSFET for the simple reason that the IGBT begins life on the production line similar to that of a conventional power MOSFET. The hybrid nature of the IGBT can be inferred by the several symbols for this device (Fig. 3-18). Standardization of the symbol can be anticipated to evolve as the device gains popularity.

The drain or collector (both terminologies for the output section of the device are encountered in the technical literature) region also starts life during the manufacturing cycle as a conventional power MOSFET. However, the doping profile of this region is modified so that both majority and minority charge-carriers are involved in current conduction. This causes the overall characteristics of the device to differ from those of the power MOSFET in the following ways:

- The forward voltage drop, $V_{CE(sat)}$, of the IGBT in the 500- to 1000-V operating range is significantly lower than that of a power MOSFET of similar voltage and current capability, and is more comparable to that of bipolar power transistors.

- The current density of the device is much greater than the unaltered power MOSFET. Thus, high-voltage and high-current ratings are easily achieved in a small device. A typical example is a 600-V, 50-A unit in a TO-220 package with a power dissipation rating of 160 W.

- The transconductance of the IGBT is much higher than that of a power MOSFET that has similar voltage and current ratings.

- No parasitic diode is used in the output circuit of the IGBT, in contrast to the ordinary power MOSFET, where the thin internal diode is sometimes useful and sometimes detrimental to circuit performance.

The IGBT: superb performance at the low end of the RF spectrum

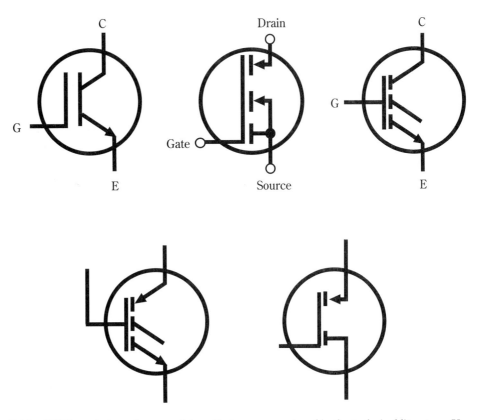

3-18 IGBT symbols and nomenclature that are encountered in the technical literature. Various semiconductor firms have also bestowed other names on this device, such as GEMFET, COMFET, MOSIGT, and IGT. The trend appears to favor IGBT because of its descriptive connotation. (Insulated-Gate Bipolar Transistor).

- The input capacitance is less than in the power MOSFET because the altered output region produces less Miller effect.
- A peculiarity of the IGBT is that the output region behaves as if there was a forward-polarized pn diode in series with the drain or collector terminal. Thus, there is no conduction below 0.7 V. This generally is of negligible consequence because the IGBT's features become more pronounced in high-voltage operation, commencing at several-hundred volts, and more often exploited in the 500- to 1000-V region. This, and other behaviors can be seen in the output characteristics of the IGBT (Fig. 3-19).
- The most notable trade-off in the IGBT, with respect to the power MOSFET, is frequency capability. Whereas power MOSFETS can be made to work in the UHF region, the IGBT tends to lose its advantages if pushed much beyond 50 kHz, or so. Indeed, IGBTs intended for use in 5-kHz applications and lower, are superior to those manufactured to provide still-reasonable operation at 50 kHz.

3-19 Typical output characteristics of the IGBT. Notice the reverse-blocking behavior and the 0.7-V offset for forward conduction.

- Unlike the power MOSFET, the forward voltage-drop, already low, does not appreciably worsen with rising temperature.
- IGBTs are relatively-easy to parallel, retaining this feature from the prototype power-MOSFET structure.
- Unlike earlier models, modern IGBTs are relatively immune to latching (an operational mode that usually leads to destruction). For practical purposes, it is safe to say that if the IGBT in forced into latching by severe overload or abuse, power MOSFETS and bipolar power transistors would also have failed.
- The IGBT is resistant to damage from concentrated-energy effects, such as the secondary breakdown that plagues bipolar power transistors.
- The IGBT is easy to drive and the output remains under complete control of the gate input signal. Unlike ordinary thyristors, which also feature very high power capability, the IGBT, as with power MOSFETS and power bipolar devices, does not require commutating techniques.

4
Impedance-matching networks

WHEN IMPLEMENTING SOLID-STATE RF POWER CIRCUITS, YOU CANNOT USUALLY simply "throw in" a resonant circuit and obtain satisfactory results. RF power transistors display very low input and output impedances compared to tubes. Both the resistive and reactive components of these impedances vary considerably with frequency and with operating conditions. RF oscillators and amplifiers can be designed and constructed to provide reasonably acceptable performance only if two mutual requirements are met: 1) you must have an appropriate solid-state device, and 2) you must interface it with appropriate impedance-matching networks. There are other stumbling blocks, to be sure, but these two are paramount. Although the topic of impedance-matching networks can easily fill the pages of a voluminous text, the material covered in this chapter should be sufficient for many of the everyday situations that are encountered in solid-state RF power.

Basic considerations

Some aspects of impedance matching were already alluded to in chapter 2. The operation of the solid-state RF generator or amplifier depends greatly on its associated LC tank circuits. This chapter examines various ways in which optimum performance can be achieved via the selection of appropriate input and output circuits. Generally, "optimum performance" will simply be the most reasonable compromise of several contradictory parameters. For example, high Q is always desirable from the standpoint of harmonic rejection. Unfortunately, overall efficiency suffers as Q is raised. This is the operating Q (the "loaded" Q of the LC circuits). The Q of individual inductors and capacitors in a network should always be as high as is practically and economically feasible. The reason that efficiency decreases as the operating Q is made higher is caused by the increased losses produced by the higher circulating currents between the L and C elements.

Because of efficiency considerations, the operating Q of the output circuit is often designed to be about 10 or 12. The reasonable frequency selectivity is provided by such a tuned circuit without seriously degrading the efficiency of the amplifier. This might pose another conflict, however. The use of practically sized L and C elements might not always be within reach when you peg the operating Q in the vicinity of 10 or 12. One consequence of this is that transistor amplifiers that operate at powers much above the flea level cannot use the familiar pi output circuits that work so effectively with tube amplifiers.

On the other hand, input networks to transistors might have lower Qs, in the 2 to 6 range, for example. Even lower Qs are used when broadbanding is desirable. Low-Q input networks contribute to stability and makes tune-up and adjustment less critical. A high-Q input circuit will not contribute much to wave purity and harmonic rejection in the output circuit of class-B and class-C amplifiers. The main function of the input circuit is impedance matching. In this regard, it is necessary to match the requirements of both the driver and the driven stage. The input circuit impedance of a transistor is inherently low and it invariably contains an appreciable reactive component. The reactive component tends to be capacitive at low and medium frequencies, but is it often inductive at VHF and UHF frequencies. However, at these higher frequencies, the manufacturer often uses the "internal" reactances to provide a "built-in" impedance-matching network. Nonetheless, external network elements, such as transformers, might still be required.

General design approach

The rigorous calculation of impedance-matching networks can lead to considerable complexity and cumbersome mathematics. In practical RF power work, such involvement is, fortunately, not usually needed. Reasonable limits are necessary, however. Thereafter, you can readily optimize operation by experimenting or by providing variable or tapped elements. Such a combination of analytical and empirical approaches is justifiable. If for no other reason, you will seldom be in a position to know the values of all the necessary parameters. For example, you might assume that the antenna feedline is 50 Ω. Such an assumption is just that; in most cases there will be appreciable departure from the ideal resistive 50 Ω. Generally, the accompanying reactive component must be somehow tuned out. This is often a task for the output network. Thus, variable elements would be necessary even if great pains were taken to design a rigorously accurate network. This is especially true if, as is often the case, more than a single fixed frequency is involved.

For yet other reasons, simplified design approaches are desirable. The tolerances on transistor parameters are notoriously sloppy. This, alone, would generally necessitate some departure from rigorously designed networks. And, of course, the capacitances and inductances of the network elements generally involve tolerances on the order of ±10%. Nor should you forget the influence of stray parameters—especially at higher frequencies. Any deviation from a set operating voltage and current causes changes in the input and output impedances of a transistor. All things considered, your objective is to keep things simple. In pursuit of this philosophy, all output networks are considered to have an operating Q of 12, and all output loads will be considered to be 50 Ω and to be purely resistive. Moreover, mathematical shortcuts will be resorted to whenever the ultimate result is not drastically affected. In general, one or more provisions for tuning or adjusting the individual network elements will be required. Perhaps, in some cases, the elements can ultimately be "frozen" at fixed values after appropriate experimentation. Instead of merely citing the relevant equations, a practical example of each technique is worked out.

Basic mechanism of impedance transformation

The manner in which a simple LC circuit can act as an impedance transformer is not obvious from an initial inspection. This exceedingly useful behavior actually stems from a unique relationship between series and parallel resistance-reactance circuits. Every parallel XR circuit can be duplicated by an equivalent series circuit. What is implied by such equivalency, and what is its significance with regard to the function of impedance transformation?

Figure 4-1 is a parallel LR circuit together with a possibly equivalent series LR circuit. Equivalency is said to exist between these circuits when the impedances "seen" across their respective terminals are the same. In the general case, this condition obtains at a certain frequency and in such a way that R_p differs from R_s. For practical purposes, X_p and X_s can be considered as equal. The notion of equivalency derives from the fact that if these two circuits would be put in a "black box," the magnitude and phase angle of their impedances would be the same.

What has been stated about the LR circuits of Fig. 4-1 applies also to the CR circuits of Fig. 4-2. Here, again, a frequency can be found in which equivalency exists (or, conversely, at a given frequency, the elements can be manipulated to produce equivalency). In both the inductive and capacitive circuits, the equations defining equivalency are as follows:

$$R_p = R_s(Q^2 + 1)$$

$$X_p = X_s \frac{(Q^2 + 1)}{Q^2}$$

$$Q = \sqrt{\frac{R_p}{R_s} - 1}$$

104 *Impedance-matching networks*

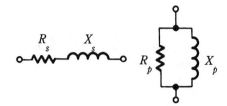

4-1 Series and parallel circuits that can be made to have similar behavior. By appropriately selecting the element values, both circuits can display the impedance magnitude and phase angle.

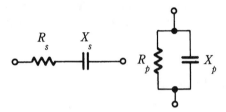

4-2 Circuits, like those of Fig. 4-1, can exhibit equivalency. Here, also, by appropriately selecting element values, the two circuits can be caused to have the same impedance level and phase angle.

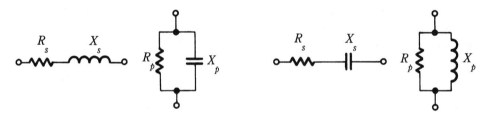

4-3 Equivalent circuits that use opposite reactances. In each of these two cases, the series circuit can be made to have the same impedance magnitude and numerical phase angle as the corresponding parallel circuit.

But in many practical situations where Q is 5 or higher, you can use the following simplifications:

$$R_p = Q^2 R_s$$
$$X_p = X_s$$

Also relevant are the equations for Q and for the specific types of reactance:

$$Q \text{ for the series circuit} = \frac{X_s}{R_s}$$

$$Q \text{ for the parallel circuit} = \frac{R_p}{X_p}$$

$$\text{Inductive reactance: } X_L = 2\pi f L$$

$$\text{Capacitive reactance: } X_C = \frac{1}{2\pi f C}$$

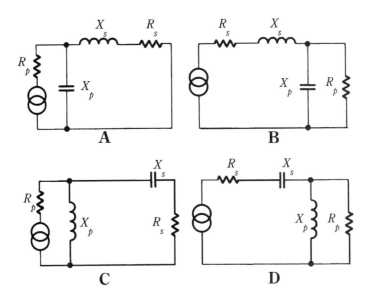

4-4 Four versions of basic "L" impedance-matching networks. All four networks are bound by relationship: $X_s/R_s = R_p/X_p$.
A. Low pass: Transforms high resistance (impedance), R_p, to lower value, R_s.
B. Low pass: Transforms low resistance (impedance), R_s, to higher value, R_p.
C. High pass: Transforms high resistance (impedance), R_p, to lower value, R_s.
D. High pass: Transforms low resistance (impedance), R_s, to higher value, R_p.

Not only can equivalences be devised from series and parallel circuits that contain one kind of reactance; they can also be devised from such circuits that contain opposite types of reactance (Fig. 4-3). As before, the series resistance (R_s) tends to be lower than the parallel resistance (R_p) when equivalency between the two circuits prevails. Having gone this far, you might ponder the consequences of connecting the series and parallel network together in order to form a single network. Providing this can be done, such a network would have the ability of transforming a low resistance, such as R_s, to a higher resistance, R_p (or, conversely, a high resistance, R_p, to a low resistance, R_s). This is an "evolved" L network. As will shortly be seen, L networks possess other features besides the mere ability to transform between resistive levels.

These concepts of L-network operation are not likely to be found in texts dealing with filters. The L section is, indeed, a true low-pass filter when configured with series and shunt arms (Fig. 4-4AB). However, the emphasis is on a different operating mode when the network is used primarily for its filtering action. In such instances, the operating frequency is below the resonant (cutoff) frequency, and the input and output resistances are both equal to the "characteristic resistance" of the filter. An example of such operation is a TVI filter that is inserted at the output of a transmitter. Whereas such filters are often pi or T networks, these networks are merely cascades of two or more basic L sections.

Conversely, the operating mode of the L network, when used for transformation between different resistance or impedance values, requires the condition of resonance between its inductive and capacitive reactances. The price paid for the ability to match different input and output resistances in such transformerlike fashion is that the network then becomes a single-frequency device. In practice, a band of frequencies can be accommodated because the operating Q is not high, as selective circuits go. The notion of viewing L networks as combined series and parallel RX circuits provides insights that are not readily attained from filter theory.

The interesting and significant aspect of equivalency is that the Qs of the parallel circuit and the series circuit are equal. This leads to the important equality, $X_s/R_s = R_p/X_p$. This relationship enables calculation of the element values. Only Q and one of these four parameters needs to be known to calculate the remaining two.

Now, look at the L network (Fig. 4-4A). This is actually an LCR circuit, but the concepts of equivalency described for the simpler circuits apply here, too. This is a resonant tank that looks like a parallel-resonant circuit at the input and a series-resonant circuit at the output. Again, postulate that the series-resonant Q and the parallel resonant Q have the same value. Mathematically, this network will "work" with a high value of R_p and a low value of R_s. Thus, at resonance, the concept of equivalency is satisfied in one physical circuit, and a high resistance, R_p is transformed down to a low resistance (R_s).

Such an L network is even better than the foregoing logic indicates, for it is not merely a transformer, but also a low-pass filter. Moreover, it can be used in inverse fashion (Fig. 4-4B), for transforming a low value of R_s to a high value of R_p. Yet another aspect of the L network is that it can "accommodate" reactance present in R_p and in R_s. It does this by absorbing such reactance in its own series and shunt arms. In so doing, the resonant condition occurs with some departure from calculated element values, but the desired transformation occurs. In practice, the amount of reactance that can be absorbed in this manner must not be stretched too far, or the readjustment to resonance can cause an appreciable change in operating Q. In other instances, resonance will not be readily attainable.

Last but not least, the L network is the basic "building block" of Pi and tee networks. As with the L network, the Pi and tee networks accomplish impedance transformation by simultaneously acting as parallel- and series-resonant tanks. They, too, can be low-pass filters. As such, their harmonic attenuation tends to be more effective than the simple two-element L network. The high-pass versions of the L-network (Fig. 4-4CD) are covered later in this chapter.

The best way to understand the impedance-matching properties of the versatile L network is with a simple example. Suppose that you desire to match a 1000-Ω source to a 100-Ω load at 10 MHz. First, notice that the Q of the network is not arbitrary; rather $Q_s = Q_p$ must be calculated by the basic equation:

$$Q = \sqrt{\frac{R_p}{R_s} - 1}$$

By substituting the stated values:

$$Q = \sqrt{\frac{1000}{100} - 1}$$

$$Q_s = Q_p = \text{network } Q = \sqrt{9} = 3$$
$$\text{then: } X_s = Q_s \times R_s = 3 \times 100 = 300 \ \Omega$$
$$\text{and: } X_p = \frac{R_p}{Q_p} = \frac{1000}{3} = 333 \ \Omega$$

Remember that one of these reactances must be inductive, and the other must be capacitive, which is generally dictated by consideration of two-factors; the low-pass configuration (inductive series arm, capacitive-shunt arm) is superior to the high-pass connection (capacitive-series arm, inductive-shunt arm) in the matter of harmonic attenuation. However, in practice, the high-pass circuit will often allow more available, less costly, or physically smaller components. In any event, both types perform the same impedance-matching function. For the low-pass versions (Fig. 4AB), the following calculations establish the actual values of the inductor and the capacitor:

$$L = \frac{X_s}{2\pi f} = \frac{300}{2\pi \times 10 \times 10^6} = 4.77 \ \mu H$$

$$C = \frac{1}{2\pi \times 10 \times 10^6 \times 333} = 48 \ pF$$

Although they are close, the values of X_p and X_s are not the same. This occurs because the behavior of parallel resonance in low-Q circuits in which ohmic resistance has a detuning effect. No such effect exists in low-Q series resonance. Although many RF impedance transformations are of the low-Q variety, a little experimentation with the equations will show that by the time Q approaches 10, X_s and X_p become nearly equal. As previously pointed out, most practical applications allow simplified calculations for Qs of five and above.

Derivation of the basic Pi network

Figure 4-5 shows a Pi network drawn so that you can see that it is composed of two cascaded L networks (of the configurations in Fig. 4-4AB). Resistances R_{s1} and R_{s2} are shown in dashed lines because they are physically nonexistent. These resistances are used during the calculation of the Pi network because each L section is dealt with individually before the two are "joined." The completed Pi network operates as if a resistance (R_{ss}) was connected across the mid-section of the Pi network. R_{ss}, of course, is the value of the parallel combination, R_{s1} and R_{s2}. That is:

$$R_{ss} = \frac{R_{s1} \times R_{s2}}{R_{s1} + R_{s2}}$$

Appropriately, R_{ss} is known as a *virtual resistance*.
Another consequence of the joining of the individually calculated L sections is

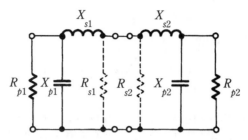

4-5 A schematic diagram that depicts a Pi network as two cascaded L networks. Although the parallel resistances, R_{s1} and R_{s2}, are physically nonexistent, they have mathematical significance during the design procedure.

that X_{s1} and X_{s2} combine to produce a single physical inductor. Notice that inductive reactance and capacitive reactance are used in the element designations. These reactances are ultimately converted to inductance and capacitance. This approach tends to systematize design procedure, which is illustrated in a forthcoming example.

Derivation of basic tee network

Figure 4-6 shows a tee network drawn so that you can see that it is composed of two cascaded L networks (of the configuration in Fig. 4-4A, B). Resistances R_{p1} and R_{p2} are shown in dashed lines because they are physically nonexistent. These resistances are used during the calculation of the tee network because each L section is dealt with individually before the two are "joined." The completed tee network operates as if a resistance (R_{pp}) was connected across the mid-section of the tee network. R_{pp}, of course, is the value of the parallel combination, R_{p1} and R_{p2}. That is:

$$R_{pp} = \frac{R_{p1} \times R_{p2}}{R_{p1} + R_{p2}}$$

Appropriately, R_{pp} is known as a virtual resistance.

Another consequence of the joining of the individually calculated L sections is that X_{p1} and X_{p2} combine to produce a single physical capacitor. The most straightforward design of the tee network uses inductors in the series arm that are not coupled to one another. It is feasible to use a single tapped inductor, but it is then necessary to bring the coupling coefficient into the calculations. Often it is then more practical to use empirical techniques. The design procedure of the tee network is illustrated in a forthcoming example.

Wave purity in transistor amplifiers

It is not as easy to obtain acceptable wave purity from a transistor amplifier as it is from its tube counterpart. Of the two devices, the transistor takes first prize for imperfect switching operation, nonlinear dynamic performance, parametric (varactor) behavior, and anomalies that involve negative resistance, discontinuities, and the like. As if this was not enough, it often displays the tube's vulnerability to UHF and VHF parasitic oscillation, together with its own unique tendency to generate

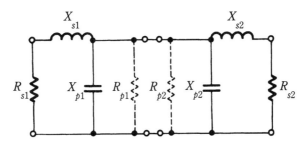

4-6 A schematic diagram tee network as two cascaded L networks. Although the parallel resistances, R_{p1} and R_{p2}, are physically nonexistent, they have mathematical significance during the design procedure.

spurious low-frequency oscillations. This follows from the fact that most RF power transistors are operated far beyond the beta cut-off frequency. This implies an increase of 6 decibels (dB) in gain for each time that the frequency is halved. It is not uncommon for the rising low-frequency gain to provoke spurious oscillations in the audio-frequency range. Most of these amplifier circuits use the common-emitter configuration. You might suppose that the low-frequency troubles could be circumvented by using the common-base circuit. However, this technique involves an undesirable trade-off: the already low input impedance of transistors is the lowest for the common-base arrangement, and new difficulties are then experienced, with respect to driver and input network design. The common-base oscillator is, however, useful at UHF and microwave frequencies.

Yet another source of frequency contamination in the output of transistor amplifiers is feedthrough from frequency-multiplier stages. This is aggravated because transistors are less effective as isolating elements than tubes. This is especially so in frequency multipliers, which, as is commonly the case, dispense with neutralizing circuits. Neutralization is generally unnecessary because of the relatively low power gain that is developed in the beta cut-off region and because the input and output circuits are tuned to different frequencies. The low gain necessitates high drive power, which, unfortunately, worsens the feedthrough of the subharmonic frequencies.

Other things being equal, the use of push-pull output stages reduces second and other even-order harmonics and thereby relaxes the filtering burden of the output network. Simply making the output network more complex often leads to awkward situations where different frequencies or antennas must be accommodated. A common practice is to use a simple output network, which functions primarily to bring about an effective impedance transformation. Then, in the interest of wave purity, a low- pass or band-pass filter is inserted in the antenna feedline.

Selection of network types for transistor RF amplifiers

The previous discussions of impedance-matching networks have been in the nature of an academic prelude to develop basic insights. This section includes practical examples of network designs. From these, the experimenter or designer can readily adapt the procedures to specific situations by merely substituting the relevant numbers. As previously stated, the emphasis is on approximate solutions via simplified equations. Experience teaches that this approach generally produces near-optimum results. Whatever modification is thereafter deemed desirable will, in most cases, involve no greater departures from the first trial design than would pertain to more rigorous computations.

Besides the basic L, Pi, and tee networks, there are many modifications of these in which certain performance parameters (such as broadbanding, tuning convenience, harmonic attenuation, etc.), are optimized. In general, the L network provides impedance transformations from high to low values, and vice versa. The tee network is particularly well suited to transform between two relatively low impedances. The Pi network is probably best adapted to transform between relatively high impedances, but as has been evident from tube amplifiers, it can also serve well in transforming between a high and a low impedance. What generally decides these matters in practice is the size, practicality, availability, and cost of the network elements. These vary greatly with application, frequency, and power level. Although it might be possible to design more than one network to comply with the requirements of a particular circuit application, the actual implementation will usually favor one over the other. This is especially true when you must also consider the voltage and current capabilities of available or readily constructed inductors and capacitors.

Except where very low power levels are involved, the requirement for impedance transformation at the output of the transistor is opposite that of tubes; that is, it is necessary to transform from a lower to the higher 50-Ω impedance of the antenna feedline. Another aspect of transistor amplifiers is the inordinately low input impedance that is presented by the base-emitter circuit (somewhat excepted from this statement is the power FET, with its tubelike input impedance). Also bearing on network design is the high input capacitance of bipolar transistors throughout the 2- to 30-MHz range and because the collector output capacitance varies with the applied voltage. The latter feature enhances harmonic production. This might be good for frequency multipliers, but it might add an extra burden to the filtering task of the output network in "straight-through amplifiers."

Some facts that pertain to all networks

Although various assumptions and approximations lead to success in most practical design approaches, one assumption generally should not be made. Do not blandly suppose that element values measured at one frequency yield the same results at some other frequency. The more widely displaced the measured frequency and the frequency involved in application, the greater is the probability of network malperformance. Also, the higher the operating frequency, the greater the likelihood that trouble will be encountered owing to element values being other than what had been measured. Lead inductance of the capacitors is one of the major contributors to this phenomenon. It is well known that every capacitor has its own resonant frequency and that at higher frequencies the capacitor begins to act like an inductor. This is readily demonstrated with a grid-dip meter and a random selection of ceramic, mica, or other capacitors. If the leads are shorted and the loop thereby formed is coupled to the grid-dip meter, resonances throughout a wide frequency range can be detected. This infers that capacitors begin to "lose" capacitance as resonance is approached from the low-frequency end of the spectrum.

In addition to the simple effect of lead inductance, more complex interplays of stray parameters can assert themselves at frequencies other than the measured frequency. This is true for inductors as well as capacitors. Indeed, inductors often exhibit alternate series and parallel resonances at frequencies that might or might not be simply related to one another. At VHF and UHF frequencies, the strays that are associated with adjacent surfaces often make it mandatory for the network to be "proved" in its actual operating environment. In any event, L and C values should be directly measured at the intended operating frequency.

The output network of a transistor power amplifier often will provide inadequate filter action for harmonics. Do not assume that a high-Q antenna is then sufficient for this task. The constants of an antenna are distributed, rather than "lumped." This permits the antenna to be responsive to frequencies other than its "resonant" frequency.

Inductors that use ferrite, powdered iron, or molybdenum cores have different inductances at different frequencies because the magnetic permeability of such substances is not constant, with respect to frequency. Also, the effective inductance of such inductors can change considerably if they are carrying direct current, or if they are operated in the region that approaches magnetic saturation.

Output impedance of bipolar transistors

As with tube amplifiers, you must know how the transistor appears to the output tank or network. To a first approximation, the transistor collector circuit displays an impedance that is determined in the same way as with tubes. That is, the approximate output impedance of a power transistor is simply $V^2/2P$, where V is the applied dc voltage and P is the anticipated power output. This approximation is valid for class-C amplifiers.

This expression is dictated by considerations for Ohm's law; it is not derived from the intrinsic characteristics of the transistor itself. It tends to give values of output resistance that err on the high side. There are two reasons for this. First, the collector saturation voltage prevents the peak values of the RF wave from actually being twice the applied dc voltage. Collector saturation voltage, V_{CE}, is often in the vicinity 1.5 V or so. Therefore, when the power supply provides 12 V, the maximum collector dc voltage is 10.5 V. Also, the relatively low-Q output networks do not quite produce the magnitude of RF peaks, which are assumed under ideal resonant-circuit behavior. All things considered, it might be best to express the output impedance of the transistor as:

$$\frac{V^2}{2P}$$

for the class-C amplifier, or as:

$$\frac{V^2}{1.57P}$$

for the class-B amplifiers. For class-A amplifiers:

$$\frac{V^2}{1.3P}$$

yields better results as approximation to output impedance.

The output transistor also has a capacitive component. It is not easy to determine this from all data sheets. For one thing, small-signal specifications or static measurements are not sufficient for the dynamic conditions that are encountered in RF power amplifiers. Actually, the output capacitance varies in varactor fashion during the excursion of the RF cycle. Therefore, the best you can do is to postulate an average value. Values within the range of 20 to 60 pF are often found. At VHF and UHF frequencies, it is often necessary to explore the effect of output capacitance in a breadboard model that closely simulates the physical layout of the final design.

Those with tube experience in RF power amplifiers can acquire the feel of solid-state versions by realizing that the resistive portion of the transistor's output impedance is the same as a hypothetical electron tube would exhibit if it could operate at several tens of volts and at plate currents in amperes.

Basic impedance-matching networks used with transistor amplifiers

The five LC networks illustrated in Fig. 4-7 represent the majority of lumped circuit impedance-matching networks that are used with transistor amplifiers operating at power levels above several watts. At lower power levels, parallel-tuned tanks with tapped or secondary windings are often encountered. In these networks, one of the impedances is considered to be 50 Ω resistive. The other impedance is considered to have a resistive and a reactive component. Although the reactive component is depicted as capacitive, inductive reactance is also encountered; for example, it is encountered at the input (base) of VHF and UHF transistors. The practical implication is that R_1 can be considered to be the output impedance of the transistor, whereupon R_L will be the idealized antenna feedline impedance. Conversely, R_1 might, in some instances, be considered to be the input impedance of the transistor, whereupon R_L will be the driving source impedance (that is, the output impedance of the driver amplifier). For sake of simplicity, first consider R_1 to be the resistive component of the output impedance of the transistor.

R_1 is always accompanied by capacitance C_{out}, which implies that the commonly used formula for output resistance:

$$R_{out} = \frac{V^2}{2P}$$

is an approximation only. Notice also that R_1 and C1 are shown as a parallel circuit in some networks and as a series circuit in others. This is not an inconsistency, but it occurs because the parallel circuit serves the purpose of computational convenience in some cases, whereas the series circuit facilitates computation in other network situations. Remember that the series and parallel circuits are convertible by the equations previously given, which are repeated in Table 4-1. Indeed, in transistor specifications, the input and output impedances of RF transistors can be given or displayed in several other ways. This might appear confusing. However, the various notational schemes all represent the same information, and any one of them can be converted to any of the others. More specifically, all can be converted into equivalent parallel

Table 4-1. Formulas applicable to the solution of the basic impedance-matching networks. When Q is 5 or greater, considerable simplification results from using Q^2, rather than $(Q^2 + 1)$ wherever this term appears. Better still, much arithmetical work can be saved by resorting to the computerized solutions listed in Table 4-2.

To convert a parallel resistance and reactance combination to series:

$$R_s = \frac{R_p}{1 + (R_p/X_p)^2}$$

$$X_s = R_s \frac{R_p}{X_p}$$

To convert a series resistance and reactance combination to parallel:

$$R_p = R_s[1 + (X_s/R_s)^2]$$

$$X_p = \frac{R_p}{X_s/R_s}$$

To solve network A (Fig. 4-7), select a Q:

$$X_{L1} = QR_1 + X_{C\,\text{out}}$$

$$X_{C2} = AR_L$$

$$X_{C1} = \frac{(B/A)(B/Q)}{(B/A) - (B/Q)} = \frac{B}{Q - A}$$

where $A = \sqrt{\left[\dfrac{R_1(1 + Q^2)}{R_L}\right] - 1}$

$B = R_1(1 + Q^2)$

To solve network B, select a Q:

$$X_{C1} = R_1/Q$$

$$X_{C2} = R_L \sqrt{\frac{R_1/R_L}{(Q^2 + 1) - (R_1/R_L)}}$$

$$X_L = \frac{QR_1 + (R_1R_L/X_{C2})}{Q^2 + 1}$$

To solve network C1, select a Q:

$$X_{L1} = X_{C\,\text{out}}$$

$$X_{C1} = QR_1$$

$$X_{C2} = R_L \sqrt{\frac{R_1}{R_L - R_1}}$$

$$X_{L2} = X_{C1} + \left(\frac{R_1 R_L}{X_{C2}}\right)$$

To solve network C2, select a Q. L_1 is not used in this network:

$$X_{C1} = QR_1$$

$$X_{C2} = R_L \sqrt{\frac{R_1}{R_L - R_1}}$$

$$X_{L2} = X_{C1} + \left(\frac{R_1 R_L}{X_{C2}}\right) + X_{C\,\text{out}}$$

To solve network D, select a Q:

$$X_{L1} = (R_1 Q) + X_{C\,\text{out}}$$

$$X_{L2} = R_L B$$

$$X_{C1} = \frac{(A/Q)(A/B)}{(A/Q) + (A/B)} = \frac{A}{Q + B}$$

where $A = R_1(1 + Q^2)$

$$B = \sqrt{\left(\frac{A}{R_L}\right) - 1}$$

---

or series RC circuits so that R_1 and C_out in the networks in Fig. 4-7 can be assigned definite values.

The reason for the additional impedance formats is that they provide advantages during the actual laboratory measurement procedure. They are covered in subsequent paragraphs.

114 Impedance-matching networks

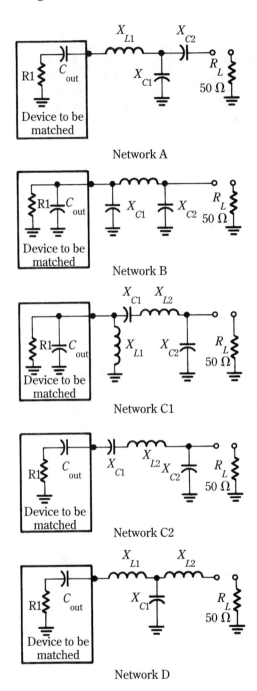

4-7 Basic impedance-matching networks that are used with transistor RF amplifiers. Convenience of computation dictates whether the device to be matched is treated as a series or a parallel circuit.

Formulas for the basic impedance-matching networks The "crank-grinding" formulas for solving the networks illustrated in Fig. 4-7 are given in Table 4-1. These formulas are predicated upon the following procedures:

- Q is selected from other considerations. Such selection is predicated on bandwidth, harmonic rejection, and the operating efficiency of the amplifier. In the real world, another governing factor enters into the selection of Q: the practicality of elements that compose the network. Inductors of a few nanohenries or of several hundred millihenries might be difficult or costly to implement. Similarly, capacitors of a few or several thousand picofarads might involve practical difficulties that are not revealed by the network formulas.
- R_1 is known and is usually assumed to be 50 Ω.
- The formulas deal with X_C, rather than C, and X_L, rather than L. This produces a generalized solution. To obtain C and L for the specific frequency (or mid-frequency) involved, use the relationships:

$$C = \frac{1}{2\pi f(X_C)} \text{ and } L = \frac{X_L}{2\pi f}.$$

- Networks A, B, and D involve the term $(Q^2 + 1)$ or $(1 + Q^2)$. Where the selected Q is 5 or greater, this term can be simplified to Q^2 in the computations. This will alleviate some of the tediousness in performing these calculations.

Computerized network solutions

The formulas listed in Table 4-1 for the determination of element values in the basic impedance-matching networks involve straightforward algebraic manipulations, but they are admittedly tedious to use. Because you will probably resort to slide rules, calculators, or tables, an even more elegant approach would be to program a digital computer to produce solutions that encompass many Q selections for all these networks. Such computerized solutions are given in Table 4-2. Here, again, select Q according to your emphasis on broadbanding, harmonic rejection, operating efficiency of the amplifier, and practicality of the elements. Some intuition pays dividends here; even the computer cannot comment on the practicality of its solutions.

These computerized solutions stem from the formulas listed in Table 4-1. Therefore, the solutions are general in that they apply for all frequencies. To arrive at a specific solution, the inductive and capacitive reactances must be converted to actual L and C values. This is done via the relationships: $L = X_L/2\pi f$ and $C = \frac{1}{2\pi f X_C}$, where f is the frequency of interest. Where considerable broadbanding exists because of low Q, f is the mid-frequency and is actually the geometric mean between the band-edge frequencies, f_1 and f_2. That is

$$f = \sqrt{f_1 \times f_2}.$$

Table 4-2. Computerized Network Solutions[a]

Network A

To design a network using the tables:

1. Transform the parallel impedance of the device to be matched to series form $(R_1 + jX_{C_{out}})$.
2. Define Q, in column one, as X_{L1}/R_1.
3. Choose a Q.
4. For a Q, find the R_s to be matched in the R column and read the reactive value of the components.
5. X_{L1} is equal to the quantity X_{L1} obtained from the tables plus $|X_{C_{out}}|'$.
6. This completes the network.

Q	X_{L1}	X_{C1}	X_{C2}	R_1
1	26	65	10	26
1	27	75.3	14.14	27
1	28	85.68	17.32	28
1	29	96.66	20	29
1	30	108.5	22.36	30
1	32	136	26.46	32
1	34	170	30	34
1	36	213.8	33.16	36
1	38	272.5	36.05	38
1	40	355	38.7	40
1	42	479	41.23	42
1	44	686.32	43.59	44
1	46	1102	45.83	46
1	48	2351	48	48
2	22	32.7	15.8	11
2	24	38.6	22.4	12
2	26	45	27.4	13
2	28	51.2	31.6	14
2	30	58	35.4	15
2	32	65.3	38.7	16
2	34	73.1	41.8	17
2	36	81.4	44.7	18
2	38	90.3	47.4	19
2	40	100	50	20
2	42	110.4	52.4	21
2	44	122	55	22
2	46	134	57	23
2	48	147	59	24
2	50	161	61	25
2	52	177	63	26
2	54	194	65	27
2	56	213	67	28
2	58	233	69	29
2	60	256	71	30
2	64	310	74	32
2	68	377	77	34
2	72	464	81	36
2	76	582	84	38
2	80	746	87	40
2	84	995	89	42
2	88	1409	92	44
2	92	2241	95	46
2	96	4739	97	48
3	18	23.5	22.3	6
3	21	29.6	31.6	7
3	24	35.9	38.7	8
3	27	42.7	44.7	9
3	30	50	50	10
3	33	57.8	54.8	11
3	36	66	59	12
3	39	75	63.2	13
3	42	84	67	14
3	45	95	71	15
3	48	105	74	16
3	51	117	77	17
3	54	130	81	18
3	57	143	84	19
3	60	158	87	20
3	63	173	89	21
3	66	190	92	22
3	69	209	95	23
3	72	228	97	24
3	75	250	100	25
3	78	274	102	26
3	81	299	105	27
3	84	327	107	28
3	87	358	110	29
3	90	393	112	30
3	96	473	116	32
3	102	575	120	34
3	108	706	124	36
3	114	882	128	38
3	120	1129	132	40
3	126	1502	136	42
3	132	2124	140	44
3	138	3372	143	46
3	144	7119	146	48
4	12	13.2	7.1	3
4	16	20	30	4
4	20	26.9	41.8	5
4	24	34.2	51	6
4	28	42.1	58.7	7
4	32	50.6	66	8
4	36	60	72	9
4	40	69	77	10
4	44	80	83	11
4	48	91	88	12
4	52	103	92	13
4	56	115	97	14
4	60	129	101	15
4	64	144	105	16
4	68	159	109	17
4	72	176	113	18
4	76	194	117	19
4	80	214	120	20
4	84	235	124	21
4	88	257	127	22
4	92	282	131	23
4	96	308	134	24
4	100	337	137	25
4	104	368	140	26
4	108	403	143	27
4	112	440	146	28
4	116	482	149	29
4	120	527	152	30
4	128	635	157	32
4	136	770	162	34
4	144	945	168	36
4	152	1180	173	38
4	160	1510	177	40
4	168	2007	182	42
4	176	2837	187	44
4	184	4500	191	46
4	192	9497	196	48
5	10	10.8	10	2
5	15	18.3	37.4	3
5	20	26.3	52	4
5	25	34.8	63.2	5
5	30	44	73	6
5	35	54	81	7
5	40	65	89	8
5	45	76	96	9
5	50	88	102	10
5	55	101	108	11
5	60	115	114	12
5	65	130	120	13
5	70	146	125	14
5	75	163	130	15
5	80	181	135	16
5	85	201	140	17
5	90	222	145	18
5	95	245	149	19
5	100	269	153	20
5	105	295	157	21
5	110	323	162	22
5	115	354	166	23
5	120	387	169	24
5	125	423	173	25
5	130	462	177	26
5	135	505	181	27
5	140	553	184	28
5	145	604	188	29
5	150	662	191	30
5	160	796	198	32
5	170	965	204	34
5	180	1184	210	36
5	190	1477	217	38
5	200	1890	222	40
5	210	2510	228	42
5	220	3548	234	44
5	230	5628	239	46
5	240	11874	245	48

([a]Motorola Semiconductor Products, Inc.)

Computerized network solutions

Q	x_{L1}	x_{C1}	x_{C2}	R_1	Q	x_{L1}	x_{C1}	x_{C2}	R_1	Q	x_{L1}	x_{C1}	x_{C2}	R_1
6	12	13.9	34.6	2	7	189	710	255	27	9	108	210	216	12
6	18	22.7	55.2	3	7	196	776	260	28	9	117	237	225	13
6	24	32.2	70	4	7	203	849	265	29	9	126	266	234	14
6	30	42.5	82	5	7	210	929	269	30	9	135	297	243	15
6	36	53.6	93	6	7	224	1117	278	32	9	144	330	251	16
6	42	65.5	102	7	7	238	1354	287	34	9	153	365	259	17
6	48	78	110	8	7	252	1661	296	36	9	162	403	267	18
6	54	92	119	9	7	266	2071	304	38	9	171	444	275	19
6	60	107	126	10	7	280	2649	312	40	9	180	488	282	20
6	66	122	133	11	7	294	3518	320	42	9	189	535	289	21
6	72	139	140	12	7	308	4971	328	44	9	198	586	296	22
6	78	157	147	13	7	322	7882	335	46	9	207	641	303	23
6	84	176	153	14	7	336	16626	343	48	9	216	701	310	24
6	90	197	159	15						9	225	766	316	25
6	96	219	165	16	8	8	8.7	27.4	1	9	234	837	323	26
6	102	242	170	17	8	16	19.3	63.2	2	9	243	914	329	27
6	108	267	175	18	8	24	31	85	3	9	252	999	335	28
6	114	295	181	19	8	32	43.6	102	4	9	261	1092	341	29
6	120	324	186	20	8	40	57.4	117	5	9	270	1196	347	30
6	126	355	191	21	8	48	72	130	6	9	288	1438	359	32
6	132	389	195	22	8	56	88	142	7	9	306	1743	370	34
6	138	426	200	23	8	64	105	153	8	9	324	2137	381	36
6	144	466	205	24	8	72	124	164	9	9	342	2665	391	38
6	150	509	209	25	8	80	143	173	10	9	360	3407	402	40
6	156	556	214	26	8	88	164	182	11	9	378	4525	412	42
6	162	608	218	27	8	96	187	191	12	9	396	6393	422	44
6	168	664	222	28	8	104	211	199	13					
6	174	727	226	29	8	112	236	207	14	10	10	11.2	50.5	1
6	180	795	230	30	8	120	264	215	15	10	20	24.5	87	2
6	192	957	238	32	8	128	293	222	16	10	30	39	112	3
6	204	1160	246	34	8	136	324	230	17	10	40	55	133	4
6	216	1422	253	36	8	144	358	237	18	10	50	72	151	5
6	228	1775	260	38	8	152	394	243	19	10	60	91	167	6
6	240	2270	267	40	8	160	433	250	20	10	70	111	181	7
6	252	3015	274	42	8	168	475	256	21	10	80	132	195	8
6	264	4260	281	44	8	176	521	263	22	10	90	155	207	9
6	276	6755	287	46	8	184	570	269	23	10	100	180	219	10
6	288	14250	294	48	8	192	623	275	24	10	110	206	230	11
					8	200	681	281	25	10	120	234	241	12
7	14	16.7	50	2	8	208	744	286	26	10	130	264	251	13
7	21	26.8	71	3	8	216	812	292	27	10	140	296	261	14
7	28	38	87	4	8	224	888	297	28	10	150	330	271	15
7	35	50	100	5	8	232	971	303	29	10	160	367	280	16
7	42	63	112	6	8	240	1062	308	30	10	170	406	289	17
7	49	77	122	7	8	256	1277	318	32	10	180	448	297	18
7	56	92	132	8	8	272	1548	329	34	10	190	494	306	19
7	63	108	141	9	8	288	1899	338	36	10	200	543	314	20
7	70	125	150	10	8	304	2368	348	38	10	210	595	322	21
7	77	143	158	11	8	320	3028	357	40	10	220	652	330	22
7	84	163	166	12	8	336	4022	366	42	10	230	713	337	23
7	91	184	173	13	8	352	5682	375	44	10	240	780	345	24
7	98	206	180	14	8	368	9009	383	46	10	250	852	352	25
7	105	230	187	15						10	260	930	359	26
7	112	256	193	16	9	9	10	40	1	10	270	1016	366	27
7	119	283	200	17	9	18	21.9	76	2	10	280	1111	373	28
7	126	313	206	18	9	27	35	99	3	10	290	1214	379	29
7	133	344	212	19	9	36	49.4	118	4	10	300	1329	383	30
7	140	379	218	20	9	45	65	134	5	10	320	1598	399	32
7	147	415	224	21	9	54	82	149	6	10	340	1937	411	34
7	154	455	229	22	9	63	100	162	7	10	360	2375	423	36
7	161	498	234	23	9	72	119	174	8	10	380	2961	435	38
7	168	544	239	24	9	81	139	185	9	10	400	3787	446	40
7	175	595	245	25	9	90	162	196	10	10	420	5029	458	42
7	182	650	250	26	9	99	185	206	11	10	440	7104	469	44

Table 4-2. Continued.

Network B

The following is a computer solution for the Pi network when R_L equals 50 ohms.

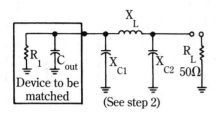

(See step 2)

To design a network using the tables:

1. Define Q, in column one, as R_1/X_{C1}.
2. C_1 actual is equal to C_1 − parallel C_{out} of device to be matched.
3. This completes the network.

Q	X_{C1}	X_{C2}	X_L	R_1	Q	X_{C1}	X_{C2}	X_L	R_1	Q	X_{C1}	X_{C2}	X_L	R_1
1	1	5.03	5.47	1	3	0.33	2.24	2.53	1	5	20	14.43	32.55	100
1	2	7.14	8	2	3	0.67	3.17	3.76	2	5	25	16.31	38.78	125
1	3	8.79	10.03	3	3	1	3.88	4.76	3	5	30	18.06	44.82	150
1	4	10.21	11.8	4	3	1.33	4.49	5.65	4	5	35	19.72	50.72	175
1	5	11.47	13.4	5	3	1.67	5.03	6.47	5	5	40	21.32	56.5	200
1	10	16.67	20	10	3	3.33	7.14	10	10	5	45	22.87	62.18	225
1	15	21	25.35	15	3	5	8.79	13.03	15	5	50	24.4	67.78	250
1	20	25	30	20	3	6.67	10.21	15.8	20	5	60	27.39	78.76	300
1	25	28.87	34.15	25	3	8.33	11.47	18.4	25	5	80	33.33	100	400
1	30	32.73	37.91	30	3	10	12.63	20.87	30	5	100	39.53	120.48	500
1	35	36.69	41.35	35	3	11.67	13.72	23.26	35	5	120	46.29	140.31	600
1	40	40.82	44.49	40	3	13.33	14.74	25.56	40	5	140	54.01	159.54	700
1	45	45.23	47.37	45	3	15	15.72	27.81	45	5	160	63.25	178.17	800
1	50	50	50	50	3	16.67	16.67	30	50	5	180	75	196.15	900
1	55	55.28	52.37	55	3	18.33	17.58	32.14	55	5	200	91.29	213.37	1000
1	60	61.24	54.49	60	3	20	18.46	34.25	60	5	220	117.26	229.58	1100
1	65	68.14	56.35	65	3	21.67	19.33	36.32	65	5	240	173.21	244.09	1200
1	70	76.38	57.91	70	3	23.33	20.17	38.35	70					
1	75	86.6	59.15	75	3	25	21	40.35	75	6	0.17	1.16	1.32	1
1	80	100	60	80	3	26.67	21.82	42.33	80	6	4.17	5.85	9.83	25
1	85	119.02	60.35	85	3	28.33	22.63	44.28	85	6	8.33	8.33	16.22	50
1	90	150	60	90	3	30	23.43	46.21	90	6	12.5	10.28	22.02	75
2	0.5	3.17	3.56	1	3	31.67	24.22	48.12	95	6	16.67	11.95	27.52	100
2	1	4.49	5.25	2	3	33.33	25	50	100	6	20.83	13.46	32.82	125
2	1.5	5.51	6.64	3	3	41.67	28.87	59.12	125	6	25	14.85	37.97	150
2	2	6.38	7.87	4	3	50	32.73	67.91	150	6	29.17	16.16	43.01	175
2	2.5	7.14	9	5	3	58.33	36.69	76.35	175	6	33.33	17.41	47.96	200
2	5	10.21	13.8	10	3	66.67	40.82	84.49	200	6	37.5	18.61	52.83	225
2	7.5	12.63	17.87	15	3	75	45.23	92.37	225	6	41.67	19.76	57.63	250
2	10	14.74	21.56	20	3	83.33	50	100	250	6	50	22	67.08	300
2	12.5	16.67	25	25						6	66.67	26.26	85.45	400
2	15	18.46	28.25	30	4	6.25	8.7	14.33	25	6	83.33	30.43	103.29	500
2	17.5	20.17	31.35	35	4	12.5	12.5	23.53	50	6	100	34.64	120.7	600
2	20	21.82	34.33	40	4	18.75	15.55	31.83	75	6	116.67	39.01	137.76	700
2	22.5	23.43	37.21	45	4	25	18.26	39.64	100	6	133.33	43.64	154.5	800
2	25	25	40	50	4	31.25	20.76	47.12	125	6	150	48.67	170.94	900
2	27.5	26.55	42.71	55	4	37.5	23.15	54.36	150	6	166.67	54.23	187.08	1000
2	30	28.1	45.35	60	4	43.75	25.46	61.39	175	6	183.33	60.55	202.93	1100
2	32.5	29.64	47.93	65	4	50	27.74	68.27	200	6	200	67.94	218.46	1200
2	35	31.18	50.45	70	4	56.25	30	75	225	6	216.67	76.87	233.66	1300
2	37.5	32.73	52.91	75	4	62.5	32.27	81.61	250	6	233.33	88.19	248.48	1400
2	40	34.3	55.32	80	4	75	36.93	94.48	300	6	250	103.51	262.83	1500
2	42.5	35.89	57.69	85	4	100	47.14	119.07	400	6	266.67	126.49	276.55	1600
2	45	37.5	60	90	4	125	59.76	142.25	500	6	283.33	168.33	289.32	1700
2	47.5	39.14	62.27	95	4	150	77.46	163.96	600	6	300	300	300	1800
2	50	40.82	64.49	100	4	175	108.01	183.77	700					
2	62.5	50	75	125	4	200	200	200	800	7	0.14	1	1.14	1
2	75	61.24	84.49	150						7	3.57	5.03	8.47	25
2	87.5	76.38	92.91	175	5	0.2	1.39	1.58	1	7	7.14	7.14	14	50
2	100	100	100	200	5	5	7	11.67	25	7	10.71	8.79	19.03	75
2	112.5	150	105	225	5	10	10	19.23	50	7	14.29	10.21	23.8	100
					5	15	12.37	26.08	75	7	17.86	11.47	28.4	125

Q	x_{C1}	x_{C2}	x_L	R_1
7	21.43	12.63	32.87	150
7	25	13.72	37.26	175
7	28.57	14.74	41.56	200
7	32.14	15.72	45.81	225
7	35.71	16.67	50	250
7	42.86	18.46	58.25	300
7	57.14	21.82	74.33	400
7	71.43	25	90	500
7	85.71	28.1	105.35	600
7	100	31.18	120.45	700
7	114.29	34.3	135.32	800
7	128.57	37.5	150	900
7	142.86	40.82	164.49	1000
7	171.43	48.04	192.98	1200
7	200	56.41	220.82	1400
7	228.57	66.67	248	1600
7	257.14	80.18	274.45	1800
7	285.71	100	300	2000
7	314.29	135.4	324.25	2200
7	342.86	244.95	345.8	2400
8	0.13	0.88	1	1
8	3.13	4.4	7.45	25
8	6.25	6.25	12.31	50
8	9.38	7.68	16.74	75
8	12.5	8.91	20.94	100
8	15.63	10	25	125
8	18.75	11	28.95	150
8	21.88	11.93	32.82	175
8	25	12.8	36.63	200
8	28.13	13.64	40.38	225
8	31.25	14.43	44.09	250
8	37.5	15.94	51.4	300
8	50	18.73	65.66	400
8	62.5	21.32	79.58	500
8	75	23.79	93.25	600
8	87.5	26.2	106.71	700
8	100	28.57	120	800
8	112.5	30.94	133.14	900
8	125	33.33	146.15	1000
8	150	38.25	171.82	1200
8	175	43.5	197.07	1400
8	200	49.24	221.92	1600
8	225	55.71	246.39	1800
8	250	63.25	270.48	2000
8	275	72.37	294.15	2200
8	300	84.02	317.36	2400
9	8.33	6.83	14.93	75
9	11.11	7.91	18.69	100
9	13.89	8.87	22.32	125
9	16.67	9.74	25.85	150
9	19.44	10.56	29.31	175
9	22.22	11.32	32.72	200
9	25	12.05	36.08	225
9	27.78	12.74	39.4	250
9	33.33	14.05	45.95	300
9	44.44	16.44	58.74	400
9	55.56	18.63	71.24	500
9	66.67	20.7	83.53	600
9	77.78	22.69	95.64	700
9	88.89	24.62	107.62	800
9	100	26.52	119.48	900
9	111.11	28.4	131.23	1000
9	133.33	32.16	154.46	1200
9	155.56	36	177.37	1400
9	177.78	40	200	1600
9	200	44.23	222.37	1800
9	222.22	48.8	244.5	2000
9	244.44	53.8	266.4	2200
9	266.67	59.41	288.05	2400
10	0.1	0.7	0.8	1
10	5	5	9.9	50
10	10	7.11	16.87	100
10	15	8.75	23.34	150
10	20	10.15	29.55	200
10	25	11.41	35.6	250
10	30	12.57	41.52	300
10	40	14.66	53.11	400
10	50	16.57	64.44	500
10	60	18.36	75.58	600
10	70	20.06	86.58	700
10	80	21.69	97.46	800
10	90	23.28	108.24	900
10	100	24.85	118.94	1000
10	120	27.91	140.09	1200
10	140	30.97	161	1400
10	160	34.05	181.68	1600
10	180	37.21	202.17	1800
10	200	40.49	222.47	2000
10	220	43.93	242.61	2200
10	240	47.58	262.59	2400
12	25	10.39	34.79	300
12	33.33	12.08	44.52	400
12	41.67	13.61	54.05	500
12	50	15.02	63.43	600
12	58.33	16.35	72.7	700
12	66.67	17.61	81.87	800
12	75	18.82	90.97	900
12	83.33	20	100	1000
12	100	22.27	117.89	1200
12	116.67	24.46	135.6	1400
12	133.33	26.61	153.15	1600
12	150	28.73	170.57	1800
12	166.67	30.86	187.86	2000
12	183.33	33	205.06	2200
12	200	35.17	222.15	2400
12	216.67	37.39	239.16	2600
12	233.33	39.66	256.07	2800
12	250	42.01	272.9	3000
12	291.67	48.3	314.64	3500
12	333.33	55.47	355.9	4000
12	375	63.96	396.67	4500
12	416.67	74.54	436.92	5000
12	458.33	88.64	476.57	5500
12	500	109.54	515.44	6000
14	21.43	8.86	29.91	300
14	28.57	10.29	38.3	400
14	35.71	11.56	46.51	500
14	42.86	12.73	54.6	600
14	50	13.83	62.59	700
14	57.14	14.87	70.51	800
14	64.29	15.86	78.37	900
14	71.43	16.81	86.17	1000
14	85.71	18.62	101.63	1200
14	100	20.35	116.95	1400
14	114.29	22.02	132.15	1600
14	128.57	23.64	147.24	1800
14	142.86	25.24	162.25	2000
14	157.14	26.81	177.17	2200
14	171.43	28.38	192.02	2400
14	185.71	29.94	206.81	2600
14	200	31.51	221.54	2800
14	214.29	33.09	236.21	3000
14	250	37.12	272.66	3500
14	285.71	41.34	308.82	4000
14	321.43	45.86	344.7	4500
14	357.14	50.77	380.33	5000
14	392.86	56.22	415.69	5500
14	428.57	62.42	450.79	6000
16	18.75	7.73	26.23	300
16	25	8.96	33.59	400
16	31.25	10.06	40.8	500
16	37.5	11.07	47.9	600
16	43.75	12	54.93	700
16	50	12.88	61.89	800
16	56.25	13.72	68.79	900
16	62.5	14.52	75.65	1000
16	75	16.05	89.26	1200
16	87.5	17.48	102.74	1400
16	100	18.86	116.12	1600
16	112.5	20.18	129.42	1800
16	125	21.47	142.64	2000
16	137.5	22.73	155.8	2200
16	150	23.96	168.9	2400
16	162.5	25.18	181.95	2600
16	175	26.39	194.96	2800
16	187.5	27.59	207.92	3000
16	218.75	30.59	240.16	3500
16	250	33.61	272.18	4000
16	281.25	36.71	304.01	4500
16	312.5	39.9	335.66	5000
16	343.75	43.25	367.15	5500
16	375	46.8	398.49	6000
18	16.67	6.86	23.35	300
18	22.22	7.94	29.9	400
18	27.78	8.91	36.33	500
18	33.33	9.79	42.66	600
18	38.89	10.61	48.92	700
18	44.44	11.38	55.13	800
18	50	12.11	61.28	900
18	55.56	12.8	67.4	1000
18	66.67	14.12	79.54	1200
18	77.78	15.35	91.57	1400
18	88.89	16.52	103.51	1600
18	100	17.65	115.38	1800
18	111.11	18.73	127.2	2000
18	122.22	19.79	138.95	2200
18	133.33	20.81	150.66	2400
18	144.44	21.82	162.33	2600
18	155.56	22.81	173.96	2800
18	166.67	23.79	185.55	3000
18	194.44	26.2	214.4	3500
18	222.22	28.57	243.08	4000
18	250	30.94	271.6	4500
18	277.78	33.33	300	5000
18	305.56	35.76	328.27	5500
18	333.33	38.25	356.44	6000
20	15	6.16	21.03	300
20	20	7.13	26.94	400
20	25	8	32.73	500
20	30	8.78	38.44	600
20	35	9.51	44.09	700
20	40	10.19	49.69	800
20	45	10.84	55.24	900
20	50	11.46	60.76	1000
20	60	12.62	71.71	1200
20	70	13.7	82.57	1400
20	80	14.72	93.35	1600
20	90	15.7	104.07	1800
20	100	16.64	114.73	2000
20	110	17.55	125.35	2200
20	120	18.44	135.93	2400
20	130	19.3	146.47	2600
20	140	20.14	156.98	2800
20	150	20.97	167.46	3000
20	175	22.99	193.54	3500
20	200	24.96	219.48	4000
20	225	26.9	245.3	4500
20	250	28.82	271.01	5000
20	275	30.74	296.62	5500
20	300	32.67	322.15	6000

Table 4-2. Continued.

Network C1

The following is a computer solution for an RF matching network. This computer solution is applicable for two forms of matching networks.

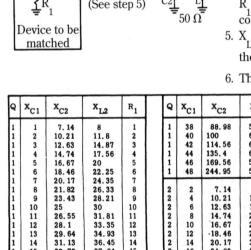

To design a network using the tables:

1. $X_{L1} = X_{Cout}$.
2. Define Q, in column one, as X_{C1}/R_1.
3. All network values can now be read from the charts in terms of reactance.
4. This completes network C_1.

Network C2

To design a network using the tables:

1. L_1 is not used in this network.
2. Transform the impedance of the device to be matched to series form $(R_1 + jX_{Cout})$.
3. Define Q, in column one, as X_{C1}/R_1.
4. For a desired Q, find the Rs to be matched in the R_1 column and read the reactive value of the components.
5. X_{L2} is equal to the quantity X_{L2} obtained from the table plus X_{Cout}.
6. This completes network C_2.

Q	X_{C1}	X_{C2}	X_{L2}	R_1	Q	X_{C1}	X_{C2}	X_{L2}	R_1	Q	X_{C1}	X_{C2}	X_{L2}	R_1
1	1	7.14	8	1	1	38	88.98	59.35	38	2	54	54.17	78.92	27
1	2	10.21	11.8	2	1	40	100	60	40	2	56	56.41	80.82	28
1	3	12.63	14.87	3	1	42	114.56	60.33	42	2	58	58.76	82.68	29
1	4	14.74	17.56	4	1	44	135.4	60.25	44	2	60	61.24	84.49	30
1	5	16.67	20	5	1	46	169.56	59.56	46	2	64	66.67	88	32
1	6	18.46	22.25	6	1	48	244.95	57.8	48	2	68	72.89	91.32	34
1	7	20.17	24.35	7						2	72	80.18	94.45	36
1	8	21.82	26.33	8	2	2	7.14	9	1	2	76	88.98	97.35	38
1	9	23.43	28.21	9	2	4	10.21	13.8	2	2	80	100	100	40
1	10	25	30	10	2	6	12.63	17.87	3	2	84	114.56	102.33	42
1	11	26.55	31.81	11	2	8	14.74	21.56	4	2	88	135.4	104.25	44
1	12	28.1	33.35	12	2	10	16.67	25	5	2	92	169.56	105.56	46
1	13	29.64	34.93	13	2	12	18.46	28.25	6	2	96	244.95	105.8	48
1	14	31.13	36.45	14	2	14	20.17	31.35	7					
1	15	32.73	37.91	15	2	16	21.82	34.33	8	3	3	7.14	10	1
1	16	34.3	39.32	16	2	18	23.43	37.21	9	3	6	10.21	15.8	2
1	17	35.89	40.69	17	2	20	25	40	10	3	9	12.63	20.87	3
1	18	37.5	42	18	2	22	26.55	42.71	11	3	12	14.74	25.56	4
1	19	39.14	43.27	19	2	24	28.1	45.35	12	3	15	16.67	30	5
1	20	40.82	44.49	20	2	26	29.64	47.93	13	3	18	18.46	34.25	6
1	21	42.55	45.68	21	2	28	31.18	50.45	14	3	21	20.17	38.35	7
1	22	44.32	46.82	22	2	30	32.73	52.91	15	3	24	21.82	42.33	8
1	23	46.15	47.92	23	2	32	34.3	55.32	16	3	27	23.43	46.21	9
1	24	48.04	48.98	24	2	34	35.89	57.69	17	3	30	25	50	10
1	25	50	50	25	2	36	37.5	60	18	3	33	26.55	53.71	11
1	26	52.04	50.98	26	2	38	39.14	62.27	19	3	36	28.1	57.35	12
1	27	54.17	51.92	27	2	40	40.82	64.49	20	3	39	29.64	60.98	13
1	28	56.41	52.82	28	2	42	42.55	66.68	21	3	42	31.18	64.45	14
1	29	58.76	53.68	29	2	44	44.32	68.82	22	3	45	32.73	67.91	15
1	30	61.24	54.49	30	2	46	46.15	70.92	23	3	48	34.3	71.32	16
1	32	66.67	56	32	2	48	48.04	72.98	24	3	51	35.89	74.69	17
1	34	72.89	57.32	34	2	50	50	75	25	3	54	37.5	78	18
1	36	80.18	58.45	36	2	52	52.04	76.98	26	3	57	39.14	81.27	19

Computerized network solutions

Q	X_{C1}	X_{C2}	X_{L2}	R_1
3	60	40.82	84.49	20
3	63	42.55	87.68	21
3	66	44.32	90.82	22
3	69	46.15	93.93	23
3	72	48.04	96.98	24
3	75	50	100	25
3	78	52.04	102.98	26
3	81	54.17	105.92	27
3	84	56.41	108.82	28
3	87	58.76	111.68	29
3	90	61.24	114.49	30
3	96	66.67	120	32
3	102	72.89	125.32	34
3	108	80.18	130.45	36
3	114	88.98	135.35	38
3	120	100	140	40
3	126	114.56	144.33	42
3	132	135.4	148.25	44
3	138	169.56	151.56	46
3	144	244.95	153.8	48
4	4	7.14	11	1
4	8	10.21	17.8	2
4	12	12.63	23.87	3
4	16	14.74	29.56	4
4	20	16.67	35	5
4	24	18.46	40.25	6
4	28	20.17	45.35	7
4	32	21.82	50.33	8
4	36	23.43	55.21	9
4	40	25	60	10
4	44	26.55	64.71	11
4	48	28.1	69.35	12
4	52	29.64	73.93	13
4	56	31.18	78.45	14
4	60	32.73	82.91	15
4	64	34.3	87.32	16
4	68	35.89	91.69	17
4	72	37.5	96	18
4	76	39.14	100.27	19
4	80	40.82	104.49	20
4	84	42.55	108.68	21
4	88	44.32	112.82	22
4	92	46.15	116.92	23
4	96	48.04	120.98	24
4	100	50	125	25
4	104	52.04	128.98	26
4	108	54.17	132.92	27
4	112	56.41	136.82	28
4	116	58.76	140.68	29
4	120	61.24	144.49	30
4	128	66.67	152	32
4	136	72.89	159.32	34
4	144	80.18	166.45	36
4	152	88.98	173.35	38
4	160	100	180	40
4	168	114.56	186.33	42
4	176	135.4	192.25	44
4	184	169.56	197.56	46
4	192	244.95	201.8	48
5	5	7.14	12	1
5	10	10.21	19.8	2
5	15	12.63	26.87	3
5	20	14.74	33.56	4
5	25	16.67	40	5
5	30	18.46	46.25	6
5	35	20.17	52.35	7
5	40	21.82	58.33	8
5	45	23.43	64.21	9
5	50	25	70	10
5	55	26.55	75.71	11
5	60	28.1	81.35	12
5	65	29.64	86.93	13
5	70	31.18	92.45	14
5	75	32.73	97.91	15
5	80	34.3	103.32	16
5	85	35.89	108.69	17
5	90	37.5	114	18
5	95	39.14	119.27	19
5	100	40.82	124.49	20
5	105	42.55	129.68	21
5	110	44.32	134.82	22
5	115	46.15	139.92	23
5	120	48.04	144.98	24
5	125	50	150	25
5	130	52.04	154.98	26
5	135	54.17	159.92	27
5	140	56.41	164.82	28
5	145	58.76	169.68	29
5	150	61.24	174.49	30
5	160	66.67	184	32
5	170	72.89	193.32	34
5	180	80.18	202.45	36
5	190	88.98	211.35	38
5	200	100	220	40
5	210	114.56	228.33	42
5	220	135.4	236.25	44
5	230	169.56	243.56	46
5	240	244.95	249.8	48
6	6	7.14	13	1
6	12	10.21	21.8	2
6	18	12.63	29.87	3
6	24	14.74	37.56	4
6	30	16.67	45	5
6	36	18.46	52.25	6
6	42	20.17	59.35	7
6	48	21.82	66.33	8
6	54	23.43	73.21	9
6	60	25	80	10
6	66	26.55	86.71	11
6	72	28.1	93.35	12
6	78	29.64	99.93	13
6	84	31.18	106.45	14
6	90	32.73	112.91	15
6	96	34.3	119.32	16
6	102	35.89	125.69	17
6	108	37.5	132	18
6	114	39.14	138.27	19
6	120	40.82	144.49	20
6	126	42.55	150.68	21
6	132	44.32	156.82	22
6	138	46.15	162.92	23
6	144	48.04	168.98	24
6	150	50	175	25
6	156	52.04	180.98	26
6	162	54.17	186.92	27
6	168	56.41	192.82	28
6	174	58.76	198.68	29
6	180	61.24	204.49	30
6	192	66.67	216	32
6	204	72.89	227.32	34
6	216	80.18	238.45	36
6	228	88.98	249.35	38
6	240	100	260	40
6	252	114.56	270.33	42
6	264	135.4	280.25	44
6	276	169.56	289.56	46
6	288	244.95	297.8	48
7	7	7.14	14	1
7	14	10.21	23.8	2
7	21	12.63	32.87	3
7	28	14.74	41.56	4
7	35	16.67	50	5
7	42	18.46	58.25	6
7	49	20.17	66.35	7
7	56	21.82	74.33	8
7	63	23.43	82.21	9
7	70	25	90	10
7	77	26.55	97.71	11
7	84	28.1	105.35	12
7	91	29.64	112.93	13
7	98	31.18	120.45	14
7	105	32.73	127.91	15
7	112	34.3	135.32	16
7	119	35.89	142.69	17
7	126	37.5	150	18
7	133	39.14	157.27	19
7	140	40.82	164.49	20
7	147	42.55	171.68	21
7	154	44.32	178.82	22
7	161	46.15	185.92	23
7	168	48.04	192.98	24
7	175	50	200	25
7	182	52.04	206.98	26
7	189	54.17	213.92	27
7	196	56.41	220.82	28
7	203	58.76	227.68	29
7	210	61.24	234.49	30
7	224	66.67	248	32
7	238	72.89	261.32	34
7	252	80.18	274.45	36
7	266	88.98	287.35	38
7	280	100	300	40
7	294	114.56	312.33	42
7	308	135.4	324.25	44
7	322	169.56	335.56	46
7	336	244.95	345.8	48
8	8	7.14	15	1
8	16	10.21	25.8	2
8	24	12.63	35.87	3
8	32	14.74	45.56	4
8	40	16.67	55	5
8	48	18.46	64.25	6
8	56	20.17	73.35	7
8	64	21.82	82.33	8
8	72	23.43	91.21	9
8	80	25	100	10
8	88	26.55	108.71	11
8	96	28.1	117.35	12
8	104	29.64	125.93	13
8	112	31.18	134.45	14
8	120	32.73	142.91	15
8	128	34.3	151.32	16
8	136	35.89	159.69	17
8	144	37.5	168	18
8	152	39.14	176.27	19
8	160	40.82	184.49	20
8	168	42.55	192.68	21
8	176	44.32	200.82	22
8	184	46.15	208.92	23
8	192	48.04	216.98	24
8	200	50	225	25
8	208	52.04	232.98	26
8	216	54.17	240.92	27
8	224	56.41	248.82	28
8	232	58.76	256.68	29
8	240	61.24	264.49	30
8	256	66.67	280	32
8	272	72.89	295.32	34
8	288	80.18	310.45	36
8	304	88.98	325.35	38

122 Impedance-matching networks

Table 4-2. Continued.

Q	X_{C1}	X_{C2}	X_{L2}	R_1	Q	X_{C1}	X_{C2}	X_{L2}	R_1	Q	X_{C1}	X_{C2}	X_{L2}	R_1
8	320	100	340	40	9	414	169.56	427.56	46	10	120	28.1	141.35	12
8	336	114.56	354.33	42	9	432	244.95	441.8	48	10	130	29.64	151.93	13
8	352	135.4	368.25	44	9	216	48.04	240.98	24	10	140	31.18	162.45	14
8	368	169.56	381.56	46	9	225	50	250	25	10	150	32.73	172.91	15
8	384	244.95	393.8	48	9	234	52.04	258.98	26	10	160	34.3	183.32	16
9	9	7.14	16	1	9	243	54.17	267.92	27	10	170	35.89	193.69	17
9	18	10.21	27.8	2	9	252	56.41	276.82	28	10	180	37.5	204	18
9	27	12.63	38.87	3	9	261	58.76	285.88	29	10	190	39.14	214.27	19
9	36	14.74	49.56	4	9	270	61.24	294.49	30	10	200	40.82	224.49	20
9	45	16.67	60	5	9	288	66.67	312	32	10	210	42.55	234.68	21
9	54	18.46	70.25	6	9	306	72.89	329.32	34	10	220	44.32	244.82	22
9	63	20.17	80.35	7	9	324	80.18	346.45	36	10	230	46.15	254.92	23
9	72	21.82	90.33	8	9	342	88.98	363.35	38	10	240	48.04	264.98	24
9	81	23.43	100.21	9	9	360	100	380	40	10	250	50	275	25
9	90	25	110	10	9	378	114.56	396.33	42	10	260	52.04	284.98	26
9	99	26.55	119.71	11	9	396	135.4	412.25	44	10	270	54.17	294.92	27
9	108	28.1	129.35	12						10	280	56.41	304.82	28
9	117	29.64	138.93	13	10	10	7.14	17	1	10	290	58.76	314.68	29
9	126	31.18	148.45	14	10	20	10.21	29.8	2	10	300	61.24	324.49	30
9	135	32.73	157.91	15	10	30	12.63	41.87	3	10	320	66.67	344	32
9	144	34.3	167.32	16	10	40	14.74	53.56	4	10	340	72.89	363.32	34
9	153	35.89	176.69	17	10	50	16.67	65	5	10	360	80.18	382.45	36
9	162	37.5	186	18	10	60	18.46	76.25	6	10	380	88.98	401.35	38
9	171	39.17	195.27	19	10	70	20.17	87.35	7	10	400	100	420	40
9	180	40.82	204.49	20	10	80	21.82	98.33	8	10	420	114.56	438.33	42
9	189	42.55	213.68	21	10	90	23.43	109.21	9	10	440	135.4	456.25	44
9	198	44.32	222.82	22	10	100	25	120	10	10	460	169.56	473.56	46
9	207	46.15	231.92	23	10	110	26.55	130.71	11	10	480	244.95	489.8	48

Network D

The following is a computer solution for an RF "Tee" matching network.

Tuning is accomplished by using a variable capacitor for C1. Variable matching can also be accomplished by increasing X_{L2} and adding an equal amount of X_C in series in the form of a variable capacitor.

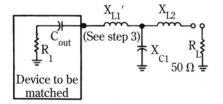

To design a network using the tables:

1. Define Q, in column one, as X_{L1}/R_1.
2. For an R_1 to be matched and a desired Q, read the reactances of the network components from the charts.
3. $X_{L1'}$ is equal to the quantity X_{L1} obtained from the tables plus $|X_{Cout}|$.
4. This completes the network.

Q	X_{L1}	X_{L2}	X_{C1}	R_1	Q	X_{L1}	X_{L2}	X_{C1}	R_1	Q	X_{L1}	X_{L2}	X_{C1}	R_1
1	26	10	43.33	26	1	175	122.47	101.46	175	2	68	77.46	47.9	34
1	27	14.14	42.09	27	1	200	132.29	109.72	200	2	72	80.62	49.83	36
1	28	17.32	41.59	28	1	225	141.42	117.54	225	2	76	83.67	51.72	38
1	29	20	41.43	29	1	250	150	125	250	2	80	86.6	53.59	40
1	30	22.36	41.46	30	1	275	158.11	132.14	275	2	84	89.44	55.43	42
1	32	26.46	41.85	32	1	300	165.83	139	300	2	88	92.2	57.23	44
1	34	30	42.5	34	2	22	15.81	23.75	11	2	92	94.87	59.01	46
1	36	33.17	43.29	36	2	24	22.36	24.52	12	2	96	97.47	60.77	48
1	38	36.06	44.16	38	2	26	27.39	25.51	13	2	100	100	62.5	50
1	40	38.73	45.08	40	2	28	31.62	26.59	14	2	110	106.07	66.73	55
1	42	41.23	46.04	42	2	30	35.36	27.7	15	2	120	111.8	70.82	60
1	44	43.59	47.01	44	2	32	38.73	28.83	16	2	130	117.26	74.8	65
1	46	45.83	48	46	2	34	41.83	29.96	17	2	140	122.47	78.66	70
1	48	47.96	49	48	2	36	44.72	31.09	18	2	150	127.48	82.43	75
1	50	50	50	50	2	38	47.43	32.22	19	2	160	132.29	86.1	80
1	55	54.77	52.49	55	2	40	50	33.33	20	2	170	136.93	89.69	85
1	60	59.16	54.96	60	2	42	52.44	34.44	21	2	180	141.42	93.2	90
1	65	63.25	57.4	65	2	44	54.77	35.54	22	2	190	145.77	96.63	95
1	70	67.08	69.79	70	2	46	57.01	36.62	23	2	200	150	100	100
1	75	70.71	62.13	75	2	48	59.16	37.7	24	2	250	169.56	115.93	125
1	80	74.16	64.43	80	2	50	61.24	38.76	25	2	300	187.08	130.62	150
1	85	77.46	66.69	85	2	52	63.25	39.82	26	2	350	203.1	144.34	175
1	90	80.62	68.9	90	2	54	65.19	40.86	27	2	400	217.94	157.26	200
1	95	83.67	71.07	95	2	56	67.08	41.9	28	2	450	231.84	169.51	225
1	100	86.6	73.21	100	2	58	68.92	42.92	29	2	500	244.95	181.19	250
1	125	100	83.33	125	2	60	70.71	43.93	30	2	550	257.39	192.37	275
1	150	111.8	92.71	150	2	64	74.16	45.93	32	2	600	269.26	203.11	300

Computerized network solutions

Q	X_{L1}	X_{L2}	X_{C1}	R_1
3	18	22.36	17.41	6
3	21	31.62	19.27	7
3	24	38.73	21.19	8
3	27	44.72	23.11	9
3	30	50	25	10
3	33	54.77	26.86	11
3	36	59.16	28.69	12
3	39	63.25	30.48	13
3	42	67.08	32.25	14
3	45	70.71	33.98	15
3	48	74.16	35.69	16
3	51	77.46	37.37	17
3	54	80.62	39.02	18
3	57	83.67	40.66	19
3	60	86.6	42.26	20
3	63	89.44	43.85	21
3	66	92.2	45.42	22
3	69	94.87	46.96	23
3	72	97.47	48.49	24
3	75	100	50	25
3	78	102.47	51.49	26
3	81	104.88	52.97	27
3	84	107.24	54.42	28
3	87	109.54	55.87	29
3	90	111.8	57.29	30
3	96	116.19	60.11	32
3	102	120.42	62.87	34
3	108	124.5	65.57	36
3	114	128.45	68.23	38
3	120	132.29	70.85	40
3	126	136.01	73.42	42
3	132	139.64	75.96	44
3	138	143.18	78.45	46
3	144	146.63	80.91	48
3	150	150	83.33	50
3	165	158.11	89.25	55
3	180	165.83	94.99	60
3	195	173.21	100.56	65
3	210	180.28	105.97	70
3	225	187.08	111.25	75
3	240	193.65	116.4	80
3	255	200	121.43	85
3	270	206.16	126.35	90
3	285	212.13	131.17	95
3	300	217.94	135.89	100
3	375	244.95	158.25	125
3	450	269.26	178.89	150
3	525	291.55	198.17	175
3	600	312.25	216.33	200
3	675	331.66	233.57	225
3	750	350	250	250
3	825	367.42	265.74	275
3	900	384.06	280.87	300
4	12	7.07	12.31	3
4	16	30	14.78	4
4	20	41.83	17.57	5
4	24	50.99	20.32	6
4	28	58.74	23	7
4	32	65.57	25.6	8
4	36	71.76	28.15	9
4	40	77.46	30.64	10
4	44	82.76	33.07	11
4	48	87.75	35.45	12
4	52	92.47	37.78	13
4	56	96.95	40.07	14
4	60	101.24	42.32	15
4	64	105.36	44.54	16
4	68	109.32	46.72	17
4	72	113.14	48.86	18
4	76	116.83	50.97	19
4	80	120.42	53.06	20
4	84	123.9	55.11	21
4	88	127.28	57.14	22
4	92	130.58	59.14	23
4	96	133.79	61.12	24
4	100	136.93	63.07	25
4	104	140	65	26
4	108	143	66.91	27

Q	X_{L1}	X_{L2}	X_{C1}	R_1
4	112	145.95	68.8	28
4	116	148.83	70.67	29
4	120	151.66	72.51	30
4	128	157.16	76.16	32
4	136	162.48	79.73	34
4	144	167.63	83.24	36
4	152	172.63	86.68	38
4	160	177.48	90.07	40
4	168	182.21	93.4	42
4	176	186.82	96.69	44
4	184	191.31	99.92	46
4	192	195.7	103.11	48
4	200	200	106.25	50
4	220	210.36	113.93	55
4	240	220.23	121.36	60
4	260	229.67	128.59	65
4	280	238.75	135.61	70
4	300	247.49	142.46	75
4	320	255.93	148.15	80
4	340	264.1	155.68	85
4	360	272.03	162.07	90
4	380	279.73	168.32	95
4	400	287.23	174.46	100
4	500	322.1	203.5	125
4	600	353.55	230.33	150
4	700	382.43	255.4	175
4	800	409.27	279.02	200
4	900	434.45	301.44	225
4	1000	458.26	322.82	250
4	1100	480.88	343.3	275
4	1200	502.49	362.99	300
5	10	10	10	2
5	15	37.42	13.57	3
5	20	51.96	17.22	4
5	25	63.25	20.75	5
5	30	72.8	24.16	6
5	35	81.24	27.47	7
5	40	88.88	30.69	8
5	45	95.92	33.82	9
5	50	102.47	36.88	10
5	55	108.63	39.87	11
5	60	114.46	42.8	12
5	65	120	45.68	13
5	70	125.3	48.49	14
5	75	130.38	51.26	15
5	80	135.28	53.99	16
5	85	140	56.67	17
5	90	144.57	59.31	18
5	95	149	61.91	19
5	100	153.3	64.47	20
5	105	157.48	67	21
5	110	161.55	69.49	22
5	115	165.53	71.96	23
5	120	169.41	74.39	24
5	125	173.21	76.79	25
5	130	176.92	79.17	26
5	135	180.55	81.52	27
5	140	184.12	83.85	28
5	145	187.62	86.15	29
5	150	191.05	88.43	30
5	160	197.74	92.91	32
5	170	204.21	97.31	34
5	180	210.48	101.63	36
5	190	216.56	105.88	38
5	200	222.49	110.06	40
5	210	228.25	114.17	42
5	220	233.88	118.21	44
5	230	239.37	122.2	46
5	240	244.74	126.13	48
5	250	260	130	50
5	275	262.68	139.46	55
5	300	274.77	148.64	60
5	325	286.36	157.54	65
5	350	297.49	166.21	70
5	375	308.22	174.66	75
5	400	318.59	182.91	80
5	425	328.63	190.97	85
5	450	338.38	198.85	90
5	475	347.85	206.57	95
5	500	357.07	214.14	100

Q	X_{L1}	X_{L2}	X_{C1}	R_1
5	625	400	250	125
5	750	438.75	283.12	150
5	875	474.34	314.08	175
5	1000	507.44	343.26	200
5	1125	538.52	670.95	225
5	1250	567.89	397.36	250
5	1375	595.82	422.67	275
5	1500	622.49	446.99	300
6	12	34.64	11.06	2
6	18	55.23	15.62	3
6	24	70	20	4
6	30	82.16	24.2	5
6	36	92.74	28.26	6
6	42	102.23	32.2	7
6	48	110.91	36.02	8
6	54	118.95	39.74	9
6	60	126.49	43.38	10
6	66	133.6	46.93	11
6	72	140.36	50.41	12
6	78	146.8	53.83	13
6	84	152.97	57.18	14
6	90	158.9	60.47	15
6	96	164.62	63.71	16
6	102	170.15	66.89	17
6	108	175.5	70.03	18
6	114	180.69	73.12	19
6	120	185.74	76.17	20
6	126	190.66	79.18	21
6	132	195.45	82.15	22
6	138	200.12	85.08	23
6	144	204.69	87.97	24
6	150	209.17	90.83	25
6	156	213.54	93.66	26
6	162	217.83	96.46	27
6	168	222.04	99.23	28
6	174	226.16	101.96	29
6	180	230.22	104.67	30
6	192	238.12	110.01	32
6	204	245.76	115.25	34
6	216	253.18	120.39	36
6	228	260.38	125.45	38
6	240	267.39	130.42	40
6	252	274.23	135.31	42
6	264	280.89	140.13	44
6	276	287.4	144.88	46
6	288	293.77	149.55	48
6	300	300	154.17	50
6	330	315.04	165.44	55
6	360	329.39	176.36	60
6	390	343.15	186.97	65
6	420	356.37	197.3	70
6	450	369.12	207.36	75
6	480	381.44	217.19	80
6	510	393.38	226.79	85
6	540	404.97	236.18	90
6	570	416.23	245.38	95
6	600	427.2	254.4	100
6	750	478.28	297.13	125
6	900	524.4	336.61	150
6	1050	566.79	373.5	175
6	1200	606.22	408.29	200
6	1350	643.23	441.3	225
6	1500	678.23	472.79	250
6	1650	711.51	502.96	275
6	1800	743.3	531.96	300
7	14	50	12.5	2
7	21	70.71	17.83	3
7	28	86.6	22.9	4
7	35	100	27.78	5
7	42	111.8	32.48	6
7	49	122.47	37.04	7
7	56	132.29	41.47	8
7	63	141.42	45.79	9
7	70	150	50	10
7	77	158.11	54.12	11
7	84	165.83	58.16	12
7	91	173.21	62.12	13
7	98	180.28	66	14
7	105	187.08	69.82	15
7	112	193.65	73.58	16

Table 4-2. Continued.

Q	X_{L1}	X_{L2}	X_{C1}	R_1
7	119	200	77.27	17
7	126	206.16	80.91	18
7	133	212.13	84.5	19
7	140	217.94	88.04	20
7	147	223.61	91.53	21
7	154	229.13	94.97	22
7	161	234.52	98.37	23
7	168	239.79	101.73	24
7	175	244.95	105.05	25
7	182	250	108.33	26
7	189	254.95	111.58	27
7	196	259.81	114.79	28
7	203	264.58	117.97	29
7	210	269.26	121.11	30
7	224	278.39	127.31	32
7	238	287.23	133.39	34
7	252	295.8	139.36	36
7	266	304.14	145.23	38
7	280	312.25	151	40
7	294	320.16	156.68	42
7	308	327.87	162.27	44
7	322	335.41	167.78	46
7	336	342.78	173.21	48
7	350	350	178.57	50
7	385	367.42	191.66	55
7	420	384.06	204.34	60
7	455	400	216.67	65
7	490	415.33	228.66	70
7	525	430.12	240.35	75
7	560	444.41	251.76	80
7	595	458.86	262.91	85
7	630	471.7	273.82	90
7	665	484.77	284.51	95
7	700	497.49	294.99	100
7	875	556.78	344.63	125
7	1050	610.33	390.49	150
7	1225	659.55	433.36	175
7	1400	705.34	473.78	200
7	1575	748.33	512.14	225
7	1750	788.99	548.73	250
7	1925	827.65	583.79	275
7	2100	864.58	617.5	300
8	8	27.39	7.6	1
8	16	63.25	14.03	2
8	24	85.15	20.1	3
8	32	102.47	25.87	4
8	40	117.26	31.42	5
8	48	130.38	36.77	6
8	56	142.3	41.95	7
8	64	153.3	46.99	8
8	72	163.55	51.9	9
8	80	173.21	56.7	10
8	88	182.35	61.39	11
8	96	191.05	65.98	12
8	104	199.37	70.49	13
8	112	207.36	74.91	14
8	120	215.06	79.26	15
8	128	222.49	83.54	16
8	136	229.67	87.74	17
8	144	236.64	91.89	18
8	152	243.41	95.97	19
8	160	250	100	20
8	168	256.42	103.97	21
8	176	262.68	107.9	22
8	184	268.79	111.77	23
8	192	274.77	115.59	24
8	200	280.62	119.38	25
8	208	286.36	123.11	26
8	216	291.98	126.81	27
8	224	297.49	130.47	28
8	232	302.9	134.09	29
8	240	308.22	137.67	30
8	256	318.59	144.73	32
8	272	328.63	151.65	34
8	288	338.38	158.46	36
8	304	347.85	165.14	38
8	320	357.07	171.71	40
8	336	366.06	178.18	42
8	352	374.83	184.56	44
8	368	383.41	190.83	46
8	384	391.79	197.02	48
8	400	400	203.13	50
8	440	419.82	218.04	55
8	480	438.75	232.49	60
8	520	456.89	246.53	65
8	560	474.34	260.2	70
8	600	491.17	273.52	75
8	640	507.44	286.52	80
8	680	523.21	299.23	85
8	720	538.52	311.66	90
8	760	553.4	323.84	95
8	800	567.89	335.78	100
8	1000	635.41	392.36	125
8	1200	696.42	444.63	150
8	1400	752.5	493.49	175
8	1600	804.67	539.57	200
8	1800	853.67	583.29	225
8	2000	900	625	250
8	2200	944.06	664.96	275
8	2400	986.15	703.38	300
9	9	40	8.37	1
9	18	75.5	15.6	2
9	27	98.99	22.4	3
9	36	117.9	28.88	4
9	45	134.16	35.09	5
9	54	148.66	41.09	6
9	63	161.86	46.91	7
9	72	174.07	52.56	8
9	81	185.47	58.07	9
9	90	196.21	63.45	10
9	99	206.4	68.71	11
9	108	216.1	73.86	12
9	117	225.39	78.92	13
9	126	234.31	83.88	14
9	135	242.9	88.76	15
9	144	251.2	93.55	16
9	153	259.23	98.28	17
9	162	267.02	102.93	18
9	171	274.59	107.51	19
9	180	281.96	112.03	20
9	189	289.14	116.49	21
9	198	296.14	120.89	22
9	207	302.99	125.23	23
9	216	309.68	129.53	24
9	225	316.23	133.77	25
9	234	322.65	137.97	26
9	243	328.94	142.12	27
9	252	335.11	146.22	28
9	261	341.17	150.28	29
9	270	347.13	154.3	30
9	288	358.75	162.23	32
9	306	370	170	34
9	324	380.92	177.63	36
9	342	391.54	185.14	38
9	360	401.87	192.52	40
9	378	411.95	199.78	42
9	396	421.78	206.93	44
9	414	431.39	213.98	46
9	432	440.79	220.93	48
9	450	450	227.78	50
9	495	472.23	244.52	55
9	540	493.46	260.74	60
9	585	513.81	276.51	65
9	630	533.39	291.85	70
9	675	552.27	306.8	75
9	720	570.53	321.4	80
9	765	588.22	335.67	85
9	810	605.39	349.63	90
9	855	622.09	363.31	95
9	900	638.36	376.71	100
9	1125	714.14	440.24	125
9	1350	782.62	498.94	150
9	1575	845.58	553.81	175
9	1800	904.16	605.54	200
9	2025	959.17	654.64	225
9	2250	1011.19	701.48	250
9	2475	1060.66	746.36	275
9	2700	1107.93	789.51	300
10	10	50.5	9.17	1
10	20	87.18	17.2	2
10	30	112.47	24.74	3
10	40	133.04	31.91	4
10	50	150.83	38.8	5
10	60	166.73	45.45	6
10	70	181.25	51.89	7
10	80	194.68	58.16	8
10	90	207.24	64.26	9
10	100	219.09	70.23	10
10	110	230.33	76.06	11
10	120	241.04	81.78	12
10	130	251.3	87.38	13
10	140	261.15	92.89	14
10	150	270.65	98.29	15
10	160	279.82	103.61	16
10	170	288.7	108.85	17
10	180	297.32	114.01	18
10	190	305.7	119.09	19
10	200	313.85	124.1	20
10	210	321.79	129.05	21
10	220	329.55	133.93	22
10	230	337.12	138.75	23
10	240	344.53	143.51	24
10	250	351.78	148.22	25
10	260	358.89	152.87	26
10	270	365.86	157.47	27
10	280	372.69	162.03	28
10	290	379.41	166.53	29
10	300	386.01	170.99	30
10	320	398.87	179.78	32
10	340	411.34	188.4	34
10	360	423.44	196.87	36
10	380	435.2	205.2	38
10	400	446.65	213.38	40
10	420	457.82	221.44	42
10	440	468.72	229.37	44
10	460	479.37	237.19	46
10	480	489.8	244.9	48
10	500	500	252.5	50
10	550	524.64	271.07	55
10	600	548.18	289.07	60
10	650	570.75	306.56	65
10	700	592.45	323.58	70
10	750	613.39	340.18	75
10	800	633.64	356.37	80
10	850	653.26	372.21	85
10	900	672.31	387.7	90
10	950	690.83	402.87	95
10	1000	708.87	417.74	100
10	1250	792.94	488.23	125
10	1500	868.91	553.36	150
10	1750	938.75	614.25	175
10	2000	1003.74	671.66	200
10	2250	1064.78	726.14	225
10	2500	1122.5	778.12	250
10	2750	1177.39	827.92	275
10	3000	1229.84	875.8	300

The bandwidth is given by f/Q and is the frequency spread between f_1 and f_2. Band edges f_1 and f_2 are actually 3 dB down, but do not be concerned with this because this condition automatically ensues from the preceding definitions.

For a given network, certain interpolations can be made to increase the usefulness of the computerized solutions. For example, if you want to design network D for a Q of 15, the X_L values would have reactances of 15/10 (1.5) times the reactances indicated for $Q = 10$. Extrapolation requires caution, and it is best to experiment first with actual indicated values before assuming that certain proportional relationships exist.

Transistor impedance information

To design impedance-matching networks, it is necessary to have at hand the output and/or the input impedance of the transistor under operating conditions that are reasonably close to those intended. This information is presented in transistor specifications sheets in several different ways, and it would be quite natural to initially experience confusion. However, the different data presentations are not the result of the manufacturer's whims—even though 100% consistency is not attained. There is more than one way to indicate impedance values. But, generally, the reasons for depicting the information in a certain format do not stem from arbitrary decisions. At different frequency ranges and under different operating conditions, the manufacturer finds it is more straightforward to use one system of measurement over another because of the very low impedances that are sometimes encountered. Also, some measuring procedures require open- or short-circuits at either the input or output of the transistor under test. This is not always feasible with RF; not only is it difficult to produce a good short- or open-circuit, but such a procedure might provoke the transistor into oscillation, and thereby ruin the measurement (and perhaps the transistor, too).

Being aware of this does not, in itself, alleviate confusion. However, remember that the various impedance presentations can all be converted into another format. Often the conversion involves simple algebraic manipulation. Where the conversion process borders on the complex, the desired readout is readily obtained from the scales on a Smith chart. In all instances, the objective is to derive either series- or parallel-equivalent impedance values so that the networks of Fig. 4-7 can be solved. (Networks A, C2, and D are shown with transistor-impedance formats that are best handled by series-equivalent impedances. The other networks are more amenable to calculation via parallel-equivalent impedances, with regard to the device to be matched. In any event, the two equivalent impedance formats are convertible by means of the first four equations listed in Table 4-1.

In Figs. 4-8 through 4-13, examples of various impedance formats are shown for randomly selected devices. The series- and parallel-equivalent impedances are illustrated, respectively, in Figs. 4-8 and 4-9. Notice that in one case, capacitive reactance is plotted, whereas in the other, capacitance, itself, is shown. Either type of plot could be used with either impedance format. Merely use the relationship: $X_C = \frac{1}{2}\pi f(C)$ or $C = \frac{1}{2} \pi f(X_C)$ to transform from capacitive reactance to capacitance, and vice versa.

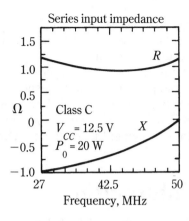 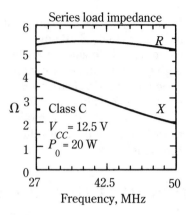

4-8 Series-equivalent impedance information. The data are readily convertible to parallel-equivalent information should the need arise. In certain networks, the series form facilitates computation of the network elements, however.

The constituents, conductance and susceptance, of admittance are shown in the graphs of Fig. 4-10. These quantities are related to the constituents of impedance, resistance, and reactance, as follows:

$$R = \frac{G}{G^2 + B^2} \quad \text{and} \quad X = \frac{B}{G^2 + B^2}$$

where:
R = resistance in ohms.
X = reactance in ohms (minus for capacitive; plus for inductive).
G = conductance in siemens.
B = susceptance in siemens (plus for capacitive; minus for inductive).

Thus, the constituents of impedance, R and X, are readily calculated from the constituents of admittance, conductance and susceptance. Although conductance corresponds to resistance and susceptance corresponds to reactance, it is not generally true that conductance is $1/R$. Such a simple reciprocal relationship exists only when the reactance is zero. Likewise, a simple reciprocal relationship between susceptance and reactance exists only when the resistance is zero. These facts can be seen from the conversion equations:

$$G = \frac{R}{R^2 + X^2} \quad \text{and} \quad B = \frac{X}{R^2 + X^2}$$

Whereas impedance, Z, obtains from $Z = \sqrt{R^2 + X^2}$, otherwise notated as $Z = R \pm jX$, admittance, Y, obtains from $Y = \sqrt{G^2 + B^2}$, otherwise notated as $Y = G \pm jB$. Admittance and impedance are connected by a simple reciprocal relationship, however. Thus, $Y = 1/Z$ and $Z = 1/Y$.

A common pitfall in the use of these entities is the fact that inductive susceptance is negative, whereas inductive reactance is positive. Capacitive susceptance is positive, whereas capacitive reactance is negative.

Admittance quantities are useful in dealing with parallel circuits because the

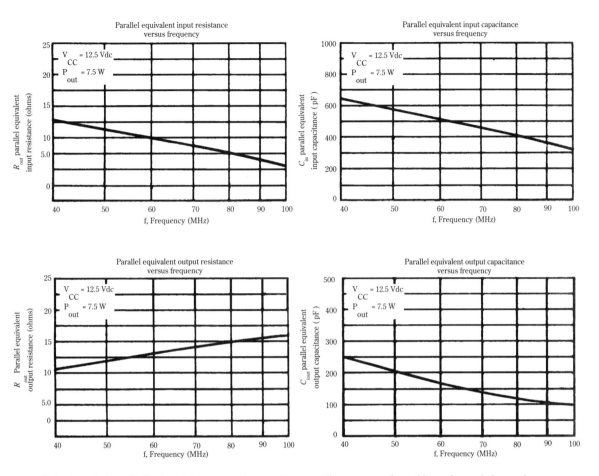

4-9 Parallel-equivalent resistances and capacitances. These examples of impedance information are suggestive of physical models of transistor input and output circuits at low to medium-high frequencies.

constituents, conductance and susceptance, combine arithmetically. Impedance quantities are useful in dealing with series circuits because the constituents, resistance and reactance, combine arithmetically. One of the beauties of the Smith chart is that both impedance and admittance quantities can be dealt with in the scaling system. It is only necessary that we be consistent when dealing in one or the other format.

A typical example of a Smith chart presentation of series-equivalent impedances is shown in Fig. 4-11. It is easy to read the resistive and reactive components for any frequency within the 200- to 500-MHz range. To help read the chart, this vendor also includes a tabular listing of the plotted resistances and reactances. Without this table, you could easily become confused in an effort to "normalize" the readings, as is done when using the Smith chart for transmission-line determinations. In this case, do not be concerned with multiplying or dividing by 50, or by other impedance values. The constituents of series impedance are simply read "as is."

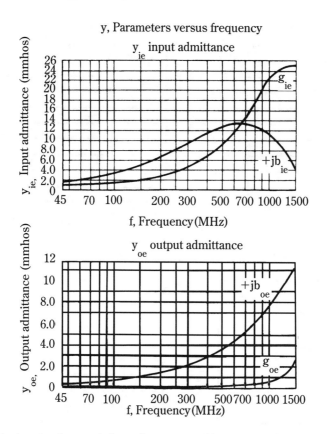

4-10 Transistor input and output information presented in terms of admittance data. In some networks, admittance data are more relevant than the corresponding impedances.

Another way to deal with the chart is to consider it normalized to 1 ohm. Then, the notation R/Z_0 simply becomes R, and the $\pm jX/Z_0$ notations (which are not seen on the limited segment of the chart shown) simply become $\pm jX$. This is tantamount to saying that the chart is read "as is." This viewpoint confirms the instructions delineated in the previous paragraph and might be a more comfortable approach for those accustomed to using the Smith chart for transmission-line problems.

Another vendor uses a graphically simplified Smith chart to display input and output impedances (Fig. 4-12). All nonrelevant scales are deleted. However, the chart is normalized to 50 Ω. Thus, all numerical values of resistance and reactance read from the chart must be multiplied by 50.

Scattering or *s* parameters

Indirect impedance information obtained from scattering parameter measuring techniques is shown in Fig. 4-13. A subsequent mathematical or graphical procedure is required to convert such a presentation to actual input and output impedance data. It is only natural to ponder the reason for displaying the impedance information in such

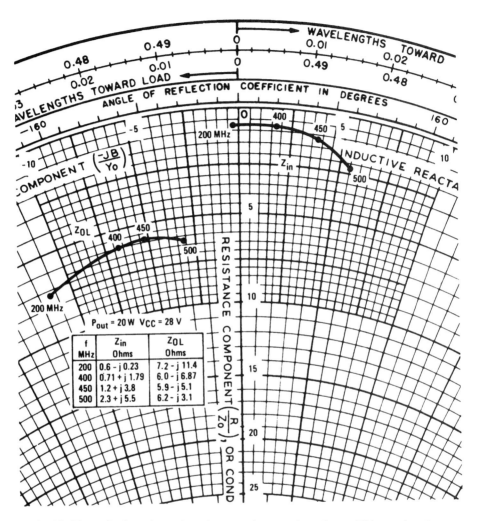

4-11 Smith Chart display of transistor input and output impedance. This vendor also presents the plotted measurements in tabular form.

indirect fashion. However, the manufacturer is able to make scattering parameter measurements more easily and more accurately than actual resistance or reactance measurements. When direct measurements of transistor input and output characteristics are made, it is necessary to use short- and open-circuits during the measurement procedure. However, such zero and infinite impedances are not easy to come by at RF frequencies—especially at VHF, UHF, and, certainly, at microwave frequencies. Even if they can be satisfactorily approximated, the transistor is often provoked into oscillation, which thereby ruins the test (and often the transistor under test).

On the other hand, when scattering parameter techniques are used, the measured quantities are voltage ratios and reflection angles. It is somewhat like evaluating the performance of a transmitter-antenna system by observing the voltage standing-wave ratio. In this measurement process, the reflection coefficient is an in-

130 *Impedance-matching networks*

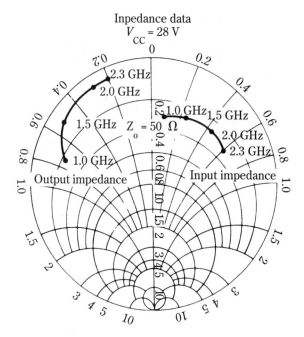

4-12 Simplified Smith Chart impedance information. All nonrelevant material has been deleted to facilitate the read-out of input and output impedances.

direct indication of impedance and can later be so converted. The significant aspect of this type of measurement is that the transistor under test is "enclosed" within 50-Ω impedances. Under this condition, oscillation or other instabilities are very unlikely, and the difficulties of approaching zero or infinite impedances are completely circumvented. Moreover, by means of 50-Ω coaxial cable, the manufacturer can locate measuring instruments remotely from the transistor being evaluated.

The apparent disadvantage of this measurement technique is that the circuit designer must incorporate extra steps in transforming the *s* parameter data to impedance information. However, if he or she already has some proficiency with Smith chart procedures, the additional work is not tedious because it can be carried out by simple graphical constructions on the Smith chart. Indeed, the designer tends to adopt the language of reflection coefficients in dealing with the input and output characteristics of transistors. Because high-frequency network elements are often transmission lines, or portions thereof, the *s* parameter display fits into the scheme of things.

However, it is not even necessary to make such graphical conversions on a Smith chart. Rather, the conversion formulas of Table 4-3 can be used. To be sure, these relationships, though straightforward, are a bit unwieldy because it is intended for the *s* parameters to be expressed as magnitudes, accompanied by phase angles; it is similarly intended that impedance parameters incorporate magnitude and phase information. In either instance, the parameters might, of course, be expressed in their rectangular-coordinate forms. For example, impedances might be expressed either as magnitude at a phase angle, or as $R \pm jX$. Such notations stem from complex algebra as commonly used in ac circuits and are not in any way peculiar to the concept of *s* parameters.

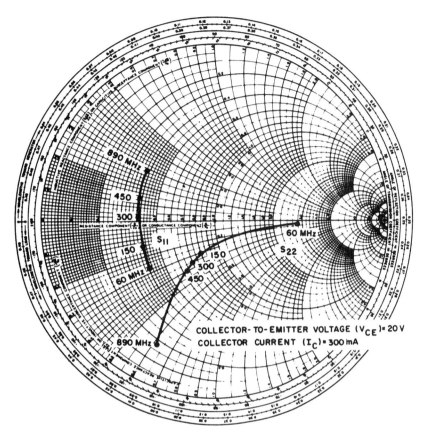

4-13 Indirect impedance display on Smith Chart from scattering parameters. Even though the manufacturer made the measurements in terms of scattering parameters, additional computations enable us to derive the corresponding input and output impedances.

Table 4-3. Conversions between *s* and *Z* parameters.
At the higher frequencies, *s*-parameter
measurement and evaluation tends to be more
practical than measuring impedances or admittances.

$$s_{11} = \frac{(Z_{11}-1)(Z_{22}+1) - Z_{12}Z_{21}}{(Z_{11}+1)(Z_{22}+1) - Z_{12}Z_{21}} \qquad Z_{11} = \frac{(1+s_{11})(1-s_{22}) + s_{12}s_{21}}{(1-s_{11})(1-s_{22}) - s_{12}s_{21}}$$

$$s_{12} = \frac{2Z_{12}}{(Z_{11}+1)(Z_{22}+1) - Z_{12}Z_{21}} \qquad Z_{12} = \frac{2s_{12}}{(1-s_{11})(1-s_{22}) - s_{12}s_{21}}$$

$$s_{21} = \frac{2Z_{21}}{(Z_{11}+1)(Z_{22}+1) - Z_{12}Z_{21}} \qquad Z_{21} = \frac{2s_{21}}{(1-s_{11})(1-s_{22}) - s_{12}s_{21}}$$

$$s_{22} = \frac{(Z_{11}+1)(Z_{22}-1) - Z_{12}Z_{21}}{(Z_{11}+1)(Z_{22}+1) - Z_{12}Z_{21}} \qquad Z_{22} = \frac{(1+s_{22})(1-s_{11}) + s_{12}s_{21}}{(1-s_{11})(1-s_{22}) - s_{12}s_{21}}$$

In practice, s parameter information is readily obtained with the use of vector voltmeters. These instruments directly provide the magnitude and phase-angle information in their read-out format.

Remember that the s parameters are reflection coefficients. Specifically, the following applies:

- s_{11} is the input reflection coefficient.
- s_{22} is the output reflection coefficient
- s_{21} is a reflection coefficient ratio that corresponds to forward transmission gain.
- s_{12} is a reflection coefficient ratio that corresponds to reverse transmission gain.

Those familiar with the use of impedance (Z) and admittance (Y) parameters will notice the correspondences in the use of the numeric subscripts.

It might be of interest to contemplate that the widely used SWR meter in amateur and CB transmitters indicates voltage SWR only because of scale calibration. In actuality, such meters respond to the ratio of reflected to incident voltage from the load (antenna). Thus, the quantity really being monitored is a reflection coefficient or s parameter. Indeed, the meter calibrations are derived from the relationship:

$$\text{SWR} = \frac{1+s}{1-s}$$

No subscripts are used with s here to avoid confusion with the actual topic discussed, the application of s parameters to active devices, such as transistors.

In addition to the relationship of s parameters to SWR measurements, simple relationships also exist for power quantities and ratios. These are often very useful in engineering design, as well as in operational appraisal. Such power relationships are listed in Table 4-4. As you can see, each is obtained from the square of an s parameter.

Table 4-4. By simply squaring the respective s parameters, additional useful performance criteria are provided.

$|s_{11}|^2 = \dfrac{\text{power reflected from the network input}}{\text{power incident on the network input}}$

$|s_{22}|^2 = \dfrac{\text{power reflected from the network output}}{\text{power incident on the network output}}$

$|s_{21}|^2 = \dfrac{\text{power delivered to a } Z_0 \text{ load}}{\text{power available from } Z_0 \text{ source}}$

$\phantom{|s_{21}|^2} = $ transducer power gain with Z_0 load and source

$|s_{12}|^2 = $ reverse transducer power gain with Z_0 load and source

Impedance matching: circuits and semantics

The basic objective of impedance matching is properly explained in engineering texts. Therein, it is explained that the maximum transfer of power to a load occurs when the load has the conjugate impedance of the circuitry that drives it. For example, if a source has an internal impedance of 10 Ω of resistance and 5 Ω of capacitive reactance, the load should have an impedance of 10 Ω of resistance and 5 Ω of inductive reactance. This is not surprising because the process is clearly one of resonance, and the net reactance of the overall circuit is canceled.

Unfortunately, many of these texts do not go beyond this idyllic, but not universally encountered, situation. For the preceding definition to be valid, you must assume that the source behaves essentially like a constant-voltage generator. The usual assumption is that the source is a perfect constant-voltage generator that is somewhat degraded from appearing as such because of internal resistance. The important point here is that this internal resistance is assumed to be relatively low.

The collector output circuit of a transistor does not comply with this assumption, however. Far from approximating a constant-voltage generator, it is a very good *constant-current generator*. This implies an extremely high internal resistance (see Fig. 4-14). It would not be practical to attempt to match to this high internal resistance. Moreover, transistor parameters are considerably different in *large-signal applications* (class C, B, and AB) than in class A.

The manufacturer rates the transistor to operate under specified peak dc voltage and power-dissipation conditions. By making an educated guess, you can postulate an expected output power from this data. Then, an Ohm's law relationship can be used to calculate an apparent internal resistance of the transistor. This is approximately given by:

$$R_L = \frac{V^2}{2(P)}$$

where:
R_L is the apparent internal resistance.
V is the dc operating voltage.
P is the estimated power that will be delivered to the load.

Then "impedance matching" is implemented between the output load impedance and this apparent transistor output impedance. Power transfer under these conditions is optimum for the conditions under which the transistor is rated, but they are not mathematically a maximum. Figure 4-14 shows the basic concept. For example, a transistor operating from a 30-Vdc supply that is intended to provide 5 W of load power would be postulated to have an apparent internal resistance of $30^2/(2 \times 5) = 90$ Ω. A transformer or matching network would then be designed to transform this value to the 50-Ω load. At the same time, conjugate reactance would be incorporated to cancel whatever reactance was "seen" at the collector of the transistor.

For the purist, this impedance-matching procedure obviously deviates from the standard definition. In actual practice, it should not be a stumbling block, however.

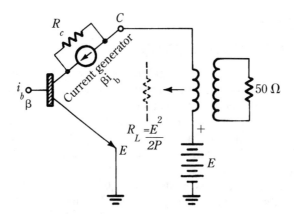

4-14 The simplified circuit showing output "impedance-matching" relationship. No attempt is made to match the high R_c resistance. Rather, the 50-Ω load resistance is transformed to the value R_L. In so doing, $\hat{P}$ is the estimated amount of power delivered to the 50-Ω load.

Both matching procedures yield optimum load power under their respective circuit and rating conditions.

A closer look at the output matching situation

What has been covered relating to "matching" the real or resistive part of the transistor's output impedance serves the purpose of achieving an approximation of proper performance. Among the simplifying assumptions made are:

- $V_{CE(sat)}$ is zero.
- There is no voltage drop in the dc collector feed system.
- The loaded Q of the output network is sufficient to produce a pure sine wave.
- All harmonics of the operating frequency "see" the zero-collector load impedance ($R_L = 0$).

None of these assumptions holds true in practical circuits. Various procedures can be resorted to in order to attain a more desirable match between the transistor and the load. For example, a more refined equation for the determination of R_L, is:

$$R_L = \frac{(E - V_{CE(sat)})^2}{2P}$$

$V_{CE(sat)}$ can be obtained from specifications sheets as a function of collector current. It will often be in the vicinity of 1.5 to 2.5 V.

Aside from these assumptions, there are other reasons why experimentation is generally desirable in order to arrive at the best output network values. There is an interplay between matching conditions for optimum power output, power gain, and transistor operating efficiency. It is generally wise to sacrifice some power gain if, in so doing, higher operating efficiency can be realized. This does not necessarily require compromise in the power output. Of course, much depends on the selection of the transistor, available drive, and the voltage regulation of the dc power source. Figure 4-15 illustrates the antagonistic relationship that is generally encountered between operating efficiency and power gain. Obviously, a relatively large improvement in efficiency can be purchased for a small reduction in power gain by making an appropriate change in the value of R_L.

Another complicating factor is that the transistor generally should not be hard-driven into saturation. Both efficiency and reliability are well served if the transistor is operated at about 80% of its maximum available power. This harks back to the all-important initial selection of the device.

The value of R_L, together with the effective parallel capacitance of the collector, compose the data for calculating the parallel equivalent output impedance of the transistor. In many cases, this capacitance is indicated in the specification sheet. It might not always be labeled clearly, but the general nomenclature "output capacitance," C_{out}, or a similar description can be taken to indicate the parallel capacitance that acts from collector to RF ground. It is qualitatively the same capacitance listed as C_{CB} for small-signal operation. Quantitatively, it is usually about twice as great as C_{CB}. In any event, the parallel-equivalent output impedance of an RF power transistor is readily computed by forming a parallel circuit of this capacitance and R_L, as determined from the operating voltage and power output.

For example, a transistor amplifier that delivers 5 W to the load and operates from a 30-V supply is computed to have an R_L value of 90 Ω. Suppose that you find in the specification sheet that this transistor has an output capacitance of 100 pF at (or in the vicinity of) the frequency that you wish to operate on. Then, the parallel-equivalent output impedance of the transistor corresponds to the impedance of 90 Ω of resistance in parallel with the reactance of the 100-pF capacitance. If the frequency of interest was 10 MHz, the reactance of the capacitance would be $\frac{1}{2\pi fC} = 159$ Ω (Fig. 4-16A).

It is often more convenient to deal with the series-equivalent circuit. To change the parallel-equivalent output impedance to the series-equivalent output impedance, and vice versa, the conversion equations shown in Table 4-1 are used. Notice that it

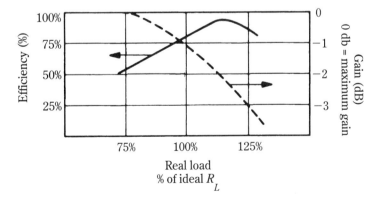

4-15 Contradictory optimization requirements for efficiency and power gain. The ideal R_L is a compromise value that trades off a small amount of power gain for an appreciable improvement in efficiency. The effect on actual power delivered to the load is not shown because of its additional dependencies on drive and thermal conditions. Generally, the desired power can be realized over a reasonable range of R_L values. Communications Transistor Corp.

136 Impedance-matching networks

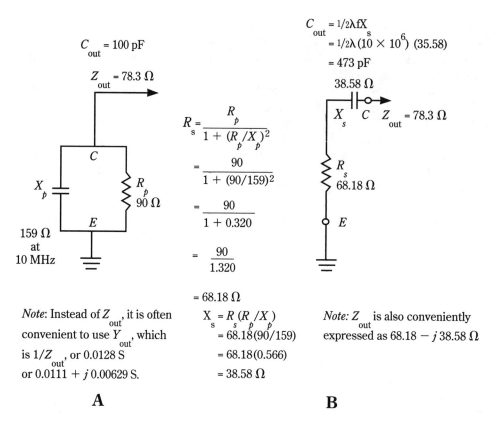

A

B

4-16 Equivalency between parallel and series impedance formats. A. Output impedance of transistor shown as a parallel RC circuit. B. Output impedance of same transistor shown as a series RC circuit. The impedance of circuit A is $1/\sqrt{(1/R_p)^2 + (1/X_p)^2} = 78.3\ \Omega$. The impedance of circuit B is $\sqrt{(R_s)^2 + (X_s)^2} = 78.3\ \Omega$.

is quite simple to convert from one circuit form to the other. Because both output impedance formats are encountered in the technical literature, such conversions tend to be a routine process when designing networks. The practicality of network element values and the ease of using the Smith chart will often dictate which circuit format is best for a given situation. Representation of the series-equivalent output impedance of the 5-W transistor amplifier is shown in Fig. 4-16B. Some of the output networks in Table 4-2 use the parallel-equivalent circuit, and others use of the series-equivalent circuit. As the term "equivalent" denotes, these circuit formats can represent the same transistor operating under similar voltage and power conditions.

Fixing the inductance of coupling, bypassing, and dc-blocking capacitors

At low RF frequencies (in the several megahertz region, for example) the selection of capacitors for coupling, bypassing, and dc blocking is often made in a manner that is suggestive of AF practice: the capacitor is decreed to be "at least large enough," and is generally chosen to be several or 10 times larger than this value. The philosophy here is that the capacitor can hardly be too large from electrical considerations because it should ideally behave as an ac short circuit. Because of the parasitic inductance inherent in all capacitors, such a design motif is bound to produce less than optimum results at higher frequencies. Indeed, the oversized capacitor can behave as a high, rather than a low impedance, such as you would infer from naively using the capacitive reactance equation: $X_C = \frac{1}{2}\pi f C$.

Whereas simulating the reactance of an ideal network capacitor had been important, the approach is somewhat different for coupling, bypass, and dc-blocking capacitors. In these functions, the objective is to make the overall reactance as low as possible, or ideally, to eliminate it altogether. To accomplish this, the capacitor is treated as a series-resonant circuit. Every practical capacitor has its natural resonant frequency because of its parasitic or lead inductance. Except for varying the lead length, you do not have control of the inductive component of the capacitor. Lead-length adjustment is sufficient for relatively low frequencies, but it is generally desirable to use "leadless" capacitors because series resonance with high inductance and relatively low capacity tends to produce high Q. This, in turn, makes the amplifier circuit more critical in adjustment and operation, and restricts bandwidth. Chip capacitors and uncased mica types are generally suitable for such self-resonant operation (Fig. 4-17).

As a consequence of the preceding considerations, the practical design procedure is to select that capacity value that resonates with its parasitic inductance at the operating frequency or at midband of the desired frequency range. Whenever possible, this resonance should be confirmed by some measurement technique or empirical test. For example, you can short the capacitor and couple a frequency dip meter to the shorting conductor. Because the short, itself, has inductance, a little skill must be developed to obtain meaningful results.

The functional difference between coupling, bypass, and dc-blocking capacitors is not always sharply defined. Indeed, sometimes such capacitors are also used to help tune the impedance-matching networks. In such instances, they would be treated as network capacitors, and the objective would not be to have them self-resonate. However, their self-inductance would still have to be accounted for to be subsequently described for network capacitors.

What to do about the inductance of network capacitors

As you use higher frequencies, the inductance in the leads and in the internal hardware of capacitors assumes ever greater importance. This is true even of chip capacitors, uncased mica capacitors, and other special RF capacitors. Elsewhere, you must be sure that the capacitor used has the proper capacitance at the actual operating frequency. The low-frequency capacitance is generally incorrect for high-frequency use. Some of this deviation can be attributed to the lump effects of measurement error, production tolerance, field fringing, radiation, proximity influences, and frequency-dependent dielectric constants. However, the parasitic inductance of the capacitor is often the most important factor involved. You cannot naively assume RF capacitors to be ideal elements. Manufacturers of RF capacitors usually indicate the value of this internal inductance under such headings as "lead," "parasitic," "internal," or "residual" inductance. Thus, you can then specify a smaller capacitor than that is calculated

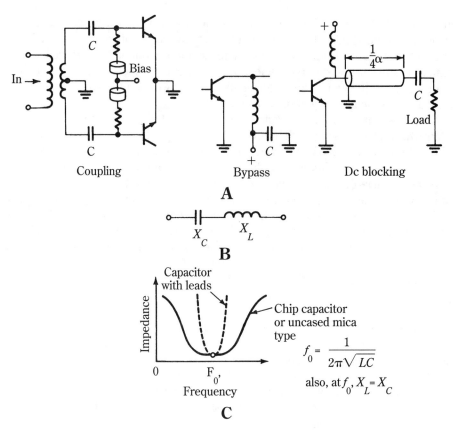

4-17 Design factors that pertain to coupling, bypass, and dc-blocking capacitors. The basic objective is to use self-resonance to attain low impedance. A. Circuit positions of coupling, bypass, and dc blocking capacitors. B. Equivalent circuit of actual capacitors, showing parasitic or lead inductance. C. Series resonance mode used for producing low impedance.

on the basis of a "pure" capacitance. The idea is to have the actual series LC circuit simulate the behavior of the originally calculated ideal capacitor (Fig. 4-18).

Circuit A has the ideal capacitor, C. Figure 4-18B shows the equivalent circuit of a real-world capacitor, with L representing the parasitic inductance. Because series inductive reactance subtracts from capacitive reactance, you must add the value of the inductive reactance to the capacitive reactance to produce the net capacitive reactance that is actually needed. Thus, the new capacitor, having higher capacitive reactance than was calculated for the ideal capacitor, has less capacitance than the original ideal capacitor. In the implementation of this logic, it is assumed that the new capacitor has the same parasitic or lead inductance as the original one. This can be a valid assumption if you stick to the same brand and type. Thus, in Fig. 4-18C, L is assumed to be the same value for new capacitor C' as for original capacitor C.

Suppose, for example, that you calculate a capacitive reactance of 10 Ω to be used as a shunt element from base to ground. You then find that the actual capaci-

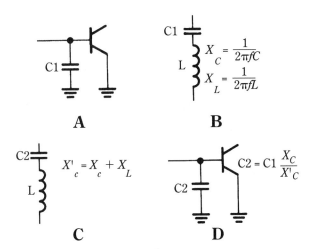

4-18 Accounting for the parasitic inductance of RF capacitors. A. Situation with ideal capacitor, C (no inductance). B. The actual electrical situation. L is the lead or parasitic inductance of capacitor. C. Recalculation of the capacitive reactance to simulate ideal capacitor. D. Implementation of new capacitor in circuit. C' is measured at a low frequency.

tor with 10 Ω of capacitive reactance also has 5 Ω of inductive reactance. The next procedure is to select a new capacitor with 15 Ω of capacitive reactance and of the same type as the originally calculated capacitor. You now have 15 Ω of capacitive reactance in series with 5 Ω of inductive reactance. This combination is the approximate equivalent of an ideal capacitor with 10 Ω of capacitive reactance. A penalty is paid in bandwidth, however, for the LC "equivalent" is frequency sensitive in the sense that it simulates the ideal capacitor only over a narrow frequency range.

Figure 4-18D uses the recalculated, or new, capacitor in circuit A. Because the parasitic inductance is not generally shown in such schematics, it is not depicted in the circuit D. Because of its presence, however, capacitor C' of circuit D is smaller than the ideal capacitor (no inductance) of circuit A. For simplicity, the transistor is not shown in B or C. The value C' is its low-frequency ("dc") value.

Use of two parallel capacitors Although almost any amount of parasitic inductance can be accounted for by this technique, the frequency sensitivity of the LC combination becomes progressively worse with higher values of inductance. In the interest of broadbanding, it is common practice to parallel two similar capacitors, rather than use a single capacitor for C'. That is, two $C''/2$ capacitances are installed. This reduces the effective parasitic inductance by half, so you only need contend with $L/2$. The use of such parallel capacitors not only makes the capacity simulation less frequency sensitive, but it halves the RF current that circulates in each capacitor. Also, a more balanced RF current path can be established in the PC board. To further enhance these features, more than two parallel capacitors are sometimes used.

The values of the two C'' capacitors are not simply one-half the value of the previously determined C' capacitor. This shouldn't be surprising, after seeing how series inductance made it necessary to alter the ideal capacitance, C, to C'. If you now reduce the effective inductance by paralleling two capacitors, the equation $X'_c = X_C + X_L$ will no longer apply to the situation. Rather, a new capacitive reactance, X''_c will be needed. The capacitive reactance per branch will then be $2X''_c$. The equation modified to apply to the new situation is $X''_c = X_C + \tfrac{1}{2} X_L$. Notice that L and X_L per capacitor are assumed the same, whether you are dealing with C, C', C'', or as will now be the case $C''/2$. This is because all these capacitors are assumed to have the same leads and very nearly the same construction features.

The logic and procedure for determining the values of the parallel capacitors are shown in Fig. 4-19. The same reasoning and assumptions involved in Fig. 4-18 apply here. Comparing step C in both procedures, the net capacitance, C'', is larger than the single capacitance, C'. If your objective is to use parallel capacitors, there is no need to calculate C'. In any event, $C''/2$ represents the low-frequency capacitance of the individual capacitors.

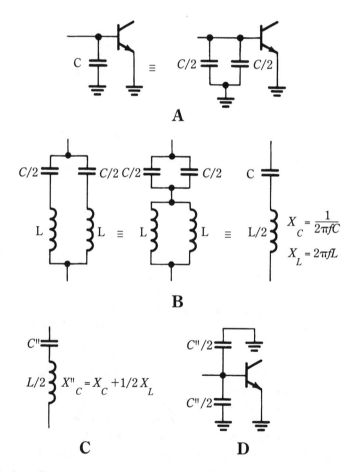

4-19 Reducing effective parasitic inductance via paralleled capacitors. A. Situation with ideal capacitors. B. The equivalent electrical situation of the parallel capacitor circuit. C. Recalculation of net capacitive reactance to simulate ideal capacitor. D. Implementation of new parallel capacitors in circuit. $C''/2$ is measured at a low frequency.

Harmonic filters

Solid-state RF amplifiers are often broadbanded and therefore are capable of supplying considerable harmonic energy to the antenna system. Even when not appreciably broadbanded, the operating Q of the output network is generally quite low in order that the transistor can develop high efficiency. This, of course, is accomplished at the expense of harmonic discrimination. Some help is provided by linear, rather than class-C operation, and push-pull circuitry materially reduces generation of even-order harmonics. Nonetheless, the practical implementation of most amplifiers requires consideration of harmonic repression ahead of the antenna. Thus, a filter must be inserted between the transmitter and the antenna feedline. Such a filter must exhibit the following features:

- Appreciable attenuation of harmonics must be provided.
- Minimal insertion loss at the fundamental frequency is essential.
- The filter must not cause excessive degradation of the VSWR.

All things considered, simple Pi and tee networks designed by easily handled image-parameter principles are generally satisfactory. Two cascaded low-pass, full sections provide a readily implemented network that will satisfactorily filter out harmonic energy in most instances. Figure 4-20 shows the pi and tee versions of such filters. At first approximation, the two configurations seem to be electrically equivalent. However, in practice, one type will generally be found to perform better than the other. You might also find that, if one end of the filter network has the shunt capacitor and the other end has the series inductor, optimum results are obtained. Which end of the filter connects to the transmitter must be determined empirically.

A basic consideration for the use of such low-pass filters is where the fundamental frequency will be allowed to fall on the response curve. The closer it is to the cutoff frequency, the more the harmonics will be attenuated. However, both insertion loss and VSWR will be adversely affected if the fundamental is too close to the cut-off frequency, which already is 3 dB down. Because of the tolerances, finite Qs, and parasitic reactances that characterize actual capacitors and inductors, some safety margin must be allowed in the design of the cutoff frequency. Generally, it is found that if the cut-off frequency is designed to be 10% higher than the fundamental, a reasonable compromise between conflicting factors is achieved.

Before attempting implementation of RF filters, the following considerations should be dealt with:

- The Qs of the inductors and capacitors should be high—in the vicinity of several hundred, at least. This should pertain in the region of the cut-off frequency. Moreover, the Qs should be no lower than about 100 at the fifth harmonic of the fundamental.
- The tolerance of inductor and capacitor values should, if feasible, be known to $\pm 5\%$. Tolerances greater than $\pm 10\%$ can readily lead to malperformance, although acceptable operation can usually be regained with "tweaking" techniques. Tolerances can be meaningless, unless they apply to the general frequency range that is embraced by the cut-off frequency of the filter.

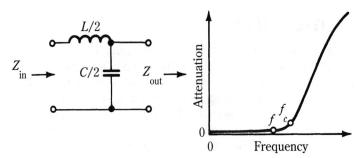

$$f_c = \frac{1}{\eta \sqrt{LC}}$$

$$Z_{in} = Z_{out} = \sqrt{L/C} = Z_0$$

$$L = \frac{Z_0}{\pi f_c}$$

$$C = \frac{1}{\pi Z_0 f_c}$$

Practical note:
f_c usually can be about 10% higher than fundamental frequency, f. If the two frequencies are too close, there will be serious insertion loss at f. If they are too far apart, harmonic attenuation will suffer.

A

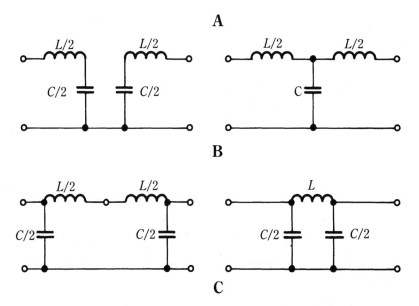

B

C

4-20 Image parameter low-pass filter. A. The L section "building block" with basic formulas and generalized response curve. B. Synthesis of a full tee section. C. Synthesis of a full pi section.

- Inductors and capacitors must be selected to have minimal parasitic reactance. This applies mostly to distributed capacitance in the inductor and to series inductance in the capacitors. In the latter case, this implies either very short leads or leadless constructions, such as capacitor "chips."

- The individual inductors in a filter must not be electromagnetically coupled (except in certain balanced configurations that are not commonly encountered). Inductors should not be too close to one another and should be oriented at right angles to each other.
- Exercise care to ensure that undesired coupling does not exist between the input and output of the filter. Such coupling can be inductive or capacitive, or can exist via radiation. Shielding techniques and the use of toroidal inductors are helpful.

Design example of a 30-MHz harmonic filter

The objective is to pass 30 MHz with minimal insertion loss and degradation of VSWR, while imparting attenuation to the second, third, and higher harmonics. It is assumed that the filter is to be inserted into a nearly flat 50-Ω antenna feedline. The design procedure is as follows:

1. From the preceding paragraphs, it appears reasonable that a two-section pi low-pass configuration could be expected to fulfill the requirements. The characteristic impedance of such a filter (Fig. 4-21) will be the same as the transmission line (50 Ω). Although more harmonic attenuation per filter element might be forthcoming from more sophisticated filter designs, this image-parameter type is exceptionally straightforward in computation and implementation.

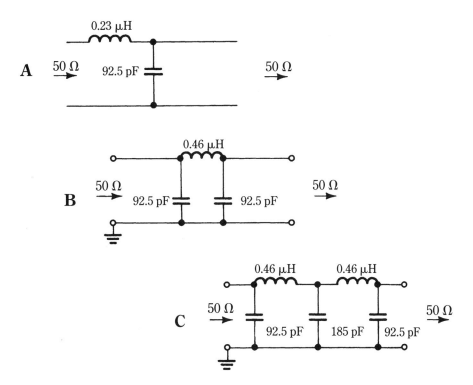

4-21 The synthesis of harmonic filter for a 30-MHz amplifier. A. A half-section building block with $Z_0 = 50\Omega$ and $f_c = 34.5$ MHz. B. Single pi section. C. Two cascaded full-pi sections.

2. Decide on the cut-off frequency, f_c. The general rule here is to make f_C about 10% or 15%, greater than the fundamental frequency. Therefore, 115% × 30 MHz = 34.5 MHz = f_C.
3. Calculate the L and C values:

a. $L = \dfrac{Z_0}{\pi f_c}$

$= \dfrac{50}{(\pi)(34.5 \times 10^6)} = 0.461\ \mu H$

and $\dfrac{L}{2} = 0.23\ \mu H$.

b. $C = \dfrac{1}{\pi Z_0 f_c}$

$= \dfrac{1}{(\pi)(50)(34.5 \times 10^6)}$

$= 0.000185\ \mu F = 185\ pF$

and $\dfrac{C}{2} = 92.5\ pF$.

4. Put the $L/2$ and $C/2$ building blocks together to form the two full-pi section filter (Fig. 4-21). Consideration should also be given to the single section of B, if the harmonic products of the amplifier are not too severe.

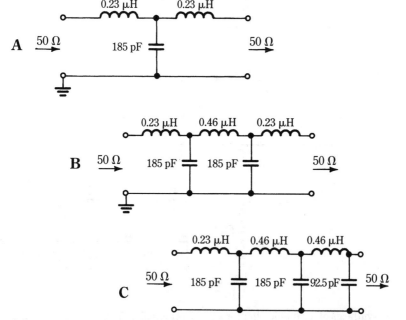

4-22 Other versions of the 30-MHz harmonic filter. A. A single tee section. B. Two cascaded full tee sections. C. Filter with one tee termination and one pi termination.

Alternative filter configurations are shown in Fig. 4-22. Notice that the tee-pi filter of Fig. 4-22C necessarily has 2.5 sections. On this basis alone, its harmonic attenuation will be better than either the two-section Pi or the two-section tee networks.

A study of these filters reveals that you cannot change a Pi input or output section to a tee section merely by omitting the end capacitor. Nor can you change a tee input or output section to a pi section merely by omitting the end inductor.

Because the fundamental frequency is not made coincident with the cut-off frequency and because of several other theoretical and practical factors, the often-quoted 6-dB/element/octave attenuation will not be realized. Approximately 4.5 to 5 decibels per element per octave is generally the practical roll-off rate. Such attenuation can often be exceeded for a specific harmonic by "tweaking" element values, but you should keep an eye on the VSWR during such experimentation.

High-pass versions of the L impedance-match network

The circuits in Fig. 4-4A, B had the configuration of low-pass filters. This stems from the use of inductors as series elements, and the use of capacitors as shunt elements. As already pointed out, an incidental benefit of this arrangement is that some attenuation of harmonics is provided. This, of course, can be beneficial in transmitter applications where it is desirable for harmonic energy to not get to the antenna.

Actually, the reactive elements of Fig. 4-4A can be interchanged; an inductor that has the same reactance (in ohms) as the capacitor can be substituted for it. Likewise, a capacitor that has the same reactance (in ohms) as the inductor can also be substituted. The important thing is that the circuit still has two opposite reactances after the described substitutions have been made. Having made the substitutions, the basic circuit configuration is no longer low-pass, but is then high-pass in nature. What has been said pertaining to Fig. 4-4A applies in similar fashion to Fig. 4-4B.

Notice carefully that this implies that you can interchange the L and C elements in a given design (providing is at least 5) and still retain the impedance-matching property calculated for the low-pass connection. This is because X_s and X_p are considered to be equal for $Q > 5$. If Q is selected or known to be less than 5, the three equations that define equivalency for inductive and capacitive circuits (page 103) should be used. The high-pass circuit versions are shown in Fig. 4-4CD. These, too, can be made into pi and tee networks.

The fact that high-pass, as well as low-pass LC circuits can be used for impedance matching was not previously discussed for the sake of simplicity, and also because the low-pass arrangements predominate in communications work. Nonetheless, high-pass networks are sometimes useful to provide a wide range of matching capability to handle different frequencies and various mismatch conditions in an antenna system. The sacrifice of harmonic attenuating performance might, furthermore, not be significant. In practice, an additional low-pass filter is needed whether the impedance matching is done by low-pass or high-pass networks. Indeed, the popular "ultimate transmatch" used in amateur radio systems is a high-pass tee network. The basic circuit is shown in Fig. 4-23. Under certain operating conditions, this network can provide a band-pass response, but it is best not to assume that harmonic attenuation will readily be attained in this way. Another version of this network is shown in Fig. 4-24. The lower half of the dual (not differential) capacitor is supposed to provide increased harmonic suppression. The claimed improvement has been a matter of controversy.

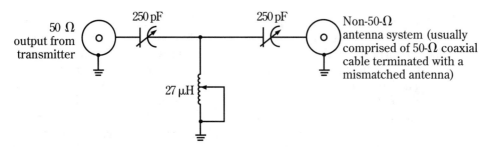

4-23 The basic circuit of the "ultimate transmatch" antenna tuner. The element values shown enable impedance matching over 3.5 to 30 MHz.

The two-control tee-circuit antenna tuner

For adjustment convenience, the shunt inductor of the high-pass tee impedance-matching network should be a continuously variable roller inductor, rather than a tapped coil. At best, because three adjustments are needed to attain or approach an SWR of 1.0 at the transmitter, initial ball-parking of low SWR values can sometimes be tricky. By the same token, tuning optimization can require a bit of patience. When changing bands, or when experimenting with the antenna system, it would be convenient to have fewer adjustments to impedance-match the transmitter to the antenna system.

Long overlooked in this type of impedance-matching network, is that the optimum impedance-match condition involves resonance of the shunt inductor by the series combination of the two capacitors, in conjunction with reactances that are absorbed from the transmitter output and from the antenna feedline. The practical implication of this statement is that one capacitor must increase as the other capacitor decreases as the resonant condition is approached. Translated into hardware, a differential capacitor can be used in place of the two separate tuning capacitors. A differential capacitor unit can be made by coupling the shafts of two separate capacitors so that their rotor plates will be mechanically displaced by 180°; while one capacitor

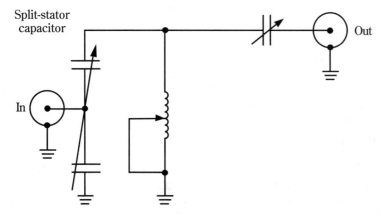

4-24 Another version of the "ultimate" or "universal" transmatch circuit. The need for the lower half of the dual capacitor is controversial.

meshes during shaft rotations, the other capacitor unmeshes. With such a provision, the impedance-match network only has two adjustments, rather than the traditional three. The circuit shown in Fig. 4-25 enables quick and easy adjustment for a wide variety of mismatch conditions throughout the 1.8- to 30-MHz range.

In communications work, these impedance-matching networks, whether of low-pass or high-pass configurations, are known as *antenna tuners*. This is somewhat of a misnomer and is sometimes responsible for commonly held confusions that pertain to the interpretation and significance of SWR readings. It would be better if these adjustable networks were described as *antenna-system tuners*. Actually, they modify the combined impedance of antenna and feedline to provide a conjugate impedance match to the transmitter. This, indeed, makes the transmitter "see" a 1:1 match. It, unfortunately, does not always ensure optimal transfer of RF power to the antenna. Also, remember that the transmitter could "see" the coveted 1:1 SWR with a dummy-load substituted for the antenna, in which case there would be no radiation of RF power. The practical significance of this statement is that the SWR monitored at the source end of a lossy feedline will be deceptively lower than you would measure at the load (antenna) end of the line. Line loss is, indeed, a "dummy" load.

The SPC transmatch impedance matcher

A relatively new member of tee-configured matching networks is the *SPC (series-parallel capacitance) transmatch* (Fig. 4-26). Its salient feature is that it is able to provide about 20-dB of harmonic rejection under virtually all source and line conditions. You might intuitively have suspected that this could be possible because of the band-pass shunt arm of the tee. However, the common notion that the SPC network is the ultimate transmatch of Fig. 4-24 used backwards is not true. A careful study of the respective circuits reveals resemblance, to be sure, but not exact identity. The unique circuit feature of the SPC network is the parallel-resonant "tank" that composes the shunt arm of the tee topography. This is not present in the reversed ultimate transmatch network.

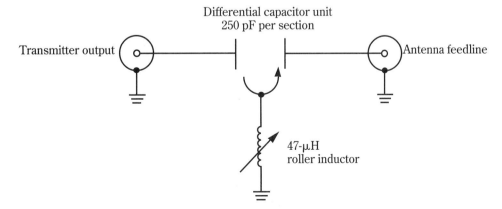

4-25 A two-control antenna tuner that uses the high-pass tee configuration. Easy adjustment over the 1.8- to 30-MHz range can be attained for widely divergent mismatch conditions.

148 Impedance-matching networks

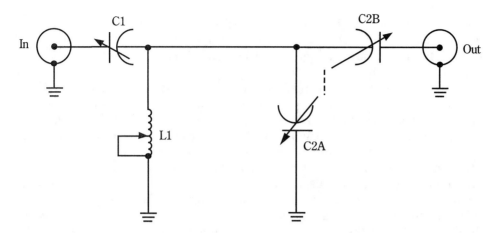

4-26 The SPC transmatch tuner. This modification of the basic tee network features enhanced harmonic suppression, and is particularly well-suited for 10-, 15-, and 20-m operation.

The SPC transmatch is particularly effective for 20, 15, 12, and 10 meters—the region of the amateur HF spectrum where the more conventional transmatches often run out of tuning range. However, to realize its full potential, good high-frequency wiring practice must be observed and the components must not be too close to the sheet metal of the cabinet. A 6-µH rotary inductor and 50-pF capacitors are sufficient for the above-indicated HF bands. Notice that the split-stator capacitor is a simple dual-section (ganged) type, not a differential type. In other words, the two-sections mesh in simultaneously. The peak voltage rating of all components depends on the power level and the VSWR that might be encountered during a tune-up adjustment. For full legal power in amateur radio work, voltage ratings in the vicinity of 4.5 to 5.0 kV are recommended. A vacuum-type variable capacitor is a worthwhile consideration for the input control, C1.

If, despite the relatively high harmonic suppression of the SPC network, it is desirable to also use a filter, the filter should be inserted between the transmitter and the SPC network. The VSWR indicator or RF wattmeter would, however, be situated between the filter and the SPC matching network. This sequential arrangement provides the best impedance environment for the filter, which is usually designed to "see" 50-Ω both ways. This advice pertains to other transmatch networks, too, assuming that they are used in the same fashion.

The Pi-L network

The high-pass tee networks can do much more than perform an impedance match between a nominal 50-Ω source and a load. Such antenna-system tuners can accommodate a wide variety of random-length single-wire antennas in both, tube and solid-state transmitters. It is only natural to ponder why such networks seem to excel low-pass circuits in ability to match widely divergent impedances. Actually, the performance disparity between the popular low-pass Pi network and the high-pass tee network might not exist in principle; what happens is that the element values often required in the low-pass Pi network become awkward or impractical. For example, a low-pass Pi network might look good on paper, only to be unrealizable when you try to implement a 2000-pF variable capacitor, with respectable voltage and current ratings. In such an instance, the high-pass tee often allows the use of readily available elements.

This being the case, it is not surprising that the so-called *Pi-L network* has become popular. This configuration (Fig. 4-27) enables the load to "see" the series arm of a tee network. Because of this, commonly encountered load impedances that would be unreachable with the basic Pi network, can be readily matched. Because the Pi-L network is low pass in nature, it is the best commonly used impedance-match network for harmonic suppression. In deciding on the type of antenna tuner to design, a key point (in practice) is that a low-reactance capacitor might be physically too large to be practical, whereas low-reactance inductors are easily and economically constructed from a few turns of wire or tubing.

Low-pass Pi and Pi-L networks often constitute the "tank" circuit of RF power amplifiers. As such, they have a moderate ability to match a variety of load situations that are presented by the transmitter end of the antenna transmission line. Sometimes, neither a low-pass harmonic filter, nor an "outboard" antenna tuner is needed.

The Pi-L network of Fig. 4-27 has found considerable use in tube final amplifiers and is particularly well suited to higher power levels. Design tables for this network are found in the ARRL *Radio Amateur's Handbook*, and in other technical literature that deals with ham-oriented topics. In principle, the tabular values pertain to a resistive load of 50 or 52 Ω. This, in practice, assumes a resonant antenna that is exactly matched to its transmission line. By the same token, such ideal operating conditions are not generally attained in practical antenna systems that are used by radio amateurs. Because the network has four variable elements, it is usually possible to compensate the departure from ideal load conditions via a tuning procedure.

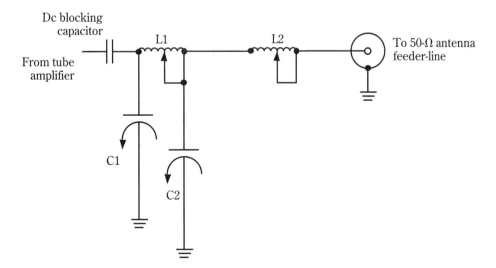

4-27 The pi-L network. C2 tends to have a more practical (lower) value than in straight pi networks. The added series inductance also confers greater harmonic suppression.

150 Impedance-matching networks

You might have questioned the wisdom of using tapped inductors that leave short-circuited coil sections in the circuit. This practice is found in other tunable networks besides the Pi-L circuit. Of course, at power-line frequencies, such a technique would be manifested by heat, smoke, and blown fuses. At RF frequencies, however, the inductive reactance of the shorted turns, plus that of the shorting path precludes any appreciable loading effect on the unshorted portion of the tapped inductor. Eddy-current dissipation from proximity effects, as well as capacitive detuning are acceptably low. As a matter of fact, because this tuning technique keeps hot RF off of the unused portion of the tapped inductor, it is a positive feature. These statements apply to air-wound coils; shorting turns on ferromagnetic toroids does not usually work out well in practice, however.

Mathematically speaking, you do not convert an extant Pi network to a Pi-L circuit by merely adding a "tail-end" series inductor of (for example), 0.5 L1. A rigorous design shows that the previous pi elements undergo change. This can be seen by comparing tables for Pi and Pi-L networks that work under the same source and load conditions. Nevertheless, experimenters often use the empirical approach of simply adding the additional output inductor to a Pi network that is known to have the desired, or near-desired impedance-match properties. This is feasible because of the four tuning provisions in the Pi-L network. This network is more often found as the tank circuit of the final amplifier than as an outboard antenna tuner. The dc-blocking capacitor should be large enough that the tuning effect on the network is negligible. High-voltage mica capacitors in the vicinity of 500 to 1000 pF are often encountered in amateur RF designs.

A unique four-element impedance-transforming network

The impedance-matching network in Fig. 4-28 bears resemblance to others dealt with, as well as to both, image-parameter filters and modern-network filter circuits.

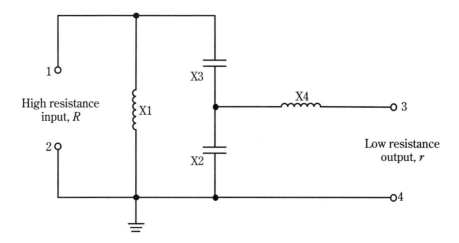

4-28 Four-element bandpass impedance transformer. One or more variable elements can serve some experimental applications, but departure from calculated values should generally be small.

Nonetheless, it is unique in design and in application. As with image-parameter filters, this network's design is predicated on the use of specific terminations. Unlike conventional image-parameter filters, this network is designed to work between dissimilar terminating resistances. Although it is true that an image-parameter filter can be modified by so-called Norton transformers to work between dissimilar terminations, this network uses a different means to accomplish the property of impedance transformation. As with image parameter and modern network filters, this network requires adherence to specific mathematical relationships between elements—proper performance cannot be realized by merely providing the required LC resonances. Its operation is suggestive of simple L networks, which simultaneously function as series and parallel resonators. Here, too, the high-resistance end of the network is the result of parallel resonance, whereas the low-resistance end is as a result of series resonance.

This network is worthy of consideration in that it can accommodate a high-impedance transformation ratio and is not critical in operation. The latter attribute is because it is basically a band-pass filter with a substantially flat passband. Thus, for many impedance matching purposes, this network can dispense with tuning adjustments. If greater matching flexibility is needed, elements X_1 and X_4 can be made variable.

If the following sequence of simple computations is followed, the design of this four-element network is quite straightforward:

1. Record the input impedance, R, and the output impedance, r, between which the network will work. Although some applications will require the input and output of Fig. 4-28 to be transposed, the higher impedance level must always be associated with terminals 1 and 2.

2. Factor: $n = \sqrt{\dfrac{R}{r}}$

3. Factor: $k = n - 1$

4. Next, you must deal with the Qs of the series and parallel resonances within the network (Q1 and Q2, respectfully). This is done by making an assumption regarding their ratio, Q_1/Q_2, for many impedance-matching needs in practical systems, a value of 2 is likely to work well. A higher value leads to a narrower passband, and a wider passband is gained from lower assumed values of Q_1/Q_2. Because your primary interest is not in the band-pass function, but rather in the impedance-transformation feature of the network, departures from the suggested value of 2 is not ordinarily a critical issue.

5. $Q_1 = \sqrt{\dfrac{Q_1}{Q_2}}(k) = \sqrt{2(k)}$ as per the previous assumption for $\dfrac{Q_1}{Q_2} = 2$

6. $Q_2 = k/Q_1$

   Now everything is in place to determine the element values:

7. $X_1 = R/Q_2$
8. $X_2 = X_1/n$
9. $X_3 = X_1 - X_2$

10. $X_4 = Q_1(r)$
11. Inductances in microhenries:

$$\frac{X(10)^6}{2\pi f}$$

where f is a geometrical center frequency in Hz

12. Capacitances in microfarads:

$$\frac{(10)^6}{2\pi f X}$$

Interestingly, the inductances and capacitances of this network can be interchanged to produce the alternative network of Fig. 4-29. Moreover, a single tapped inductor can be used for X_2 and X_3, but the design then becomes partially empirical—largely because the mutual inductance of RF coils is not always easy to evaluate. The same 12 computational steps apply to both versions of the network (shown in the illustrations).

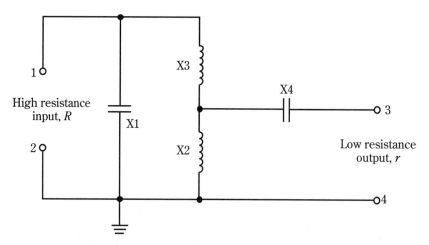

4-29 Alternate network for four-element bandpass impedance transformer. There is no mutual inductance between X2 and X3.

5
Applications of transmission-line elements to RF power circuitry

SEMICONDUCTOR DEVICES AND NETWORK THEORY HAVE BEEN COVERED AND emphasis has been placed on their interdependency in solid-state RF circuitry. Because of the ever-increasing use of the VHF, UHF, and microwave spectral regions, the impedance-matching, resonating, and filtering networks are often composed of transmission-line elements, rather than from "lumped" inductors and capacitors. Also, such elements are often better performers than conventional RF chokes or even the best bypass capacitors.

Transmission-line theory is readily available in many good textbooks, but the treatment generally serves the purposes of the mathematician, rather than the practical engineer, experimenter, or skilled hobbyist. This chapter endeavors to present this theory in so that practical implementations will obviously benefit.

Relevant, also, is the discussion of transmission-line transformers. These amazing devices represent a relatively new energy-transfer technique, one in which "primary" and "secondary" circuits are not coupled via electromagnetic induction in the manner of conventional transformers. Transmission-line transformers are destined to play very important roles in solid-state RF power. A good illustration of this is the kilowatt solid-state RF amplifier (chapter 7).

Use of transmission-line elements

Although "lumped circuit" reactances and transformers are often used in solid-state RF power systems, superior performance generally can be had with transmission-line elements. This is particularly true at VHF, UHF, and certainly at microwave frequencies, where stripline techniques contribute greatly to predictability, reproducibility, and low losses. Transmission-line devices can be used to simulate inductance, capacitance, series-, and parallel-resonant circuits, and RF chokes. They also are widely used for impedance transformation and for balun techniques (transforming from balanced to unbalanced circuitry, and vice versa). They are, additionally, often used in a variety of power splitters and power combiners. These function essentially as hybrid transformers and enable high power to be developed in the load as the summation of contributions from many amplifiers. At the same time, the individual amplifiers are electrically isolated from one another.

Transmission-line elements can take several physical forms. Parallel, coaxial, and twisted-wire lines are commonly used. Stripline elements use single lines working against a ground plane. Such lines are popular because of their compatibility with PC board techniques. Also, lines can be loaded in various ways with dielectric and ferromagnetic materials. At low frequencies, where transmission-line devices tend to be awkwardly long, they are often coiled to achieve physical compactness. Coaxial cable is, in particular, amenable to this practice, and the electrical characteristics of such a coiled line generally remain unaltered for the purpose served.

The behavior of transmission lines is governed primarily by their electrical length and by their characteristic impedance. The electrical length of practical lines is always somewhat shorter than the physical length. This is because the velocity of electromagnetic propagation along a line is less than it is in free space. Depending on the dielectric material, practical lines are often on the order of 60 to 98% of the length calculated on the basis of free-space velocity. The characteristic impedance of lines depends on constructional features and on the dielectric or ferromagnetic materials that are inserted between the lines. Characteristic impedance values between 20 and 200 Ω are often encountered, but higher and lower levels can be attained where inordinately great or small spacing can be accommodated. In any event, the impedance "seen" by the circuit terminations also depends on the manner in which the lines are used. Schematic diagrams of RF circuits are often drawn with the L and C components that the transmission-line elements simulate.

Characteristic impedance of transmission lines

In the many different uses involving transmission lines in solid-state RF circuitry, the behavior of the line is invariably governed by its characteristic impedance and its effective electrical length. Characteristic impedance is a function of the line's construction and geometry, and of the dielectric constant of the insulating medium. The effective electrical length depends primarily on the dielectric constant, but it might be also influenced by "proximity" effects from other conductors or insulators, and by fringing or distortion of the electric lines of force in the insulating medium. Finally, if any magnetically permeable substance is interposed between the lines, both the characteristic impedance and the effective electrical length will be affected. Transmission lines that have air as the dielectric medium will, ideally, have very nearly the same physical and electrical lengths. In practice, other effects generally make the physical length slightly less than the effective electrical length. Only in "free space" and under ideal conditions are the physical and electrical lengths identical. In real life, the physical length will always be less than the effective electrical length: sometimes negligibly so, as in certain air-dielectric lines; sometimes considerably so, as in lines that use other dielectric substances. The physical length of a transmission line is $1/\sqrt{e}$ times its electrical length, where e is the dielectric constant of the substance between the line elements. For air, e is very close to its "free space" value of 1. All other commonly used substances have higher dielectric constants.

From the preceding, it is apparent that the parameters of a transmission line are interrelated. Your insight into the nature of transmission-line applications to solid-state RF circuits can be enhanced by contemplating the following situations.

The characteristic impedance, Z_0, of any type or length of line can be determined by making two measurements: the input impedance of the line when the far end is open-circuited, and the input impedance of the line when the far end is short-circuited. The equation describing this relationship is: $Z_0 = \sqrt{Z_{oc} \times Z_{sc}}$, where Z_{oc} is the input impedance with the far end of the line open-circuited, and Z_{sc} is the input impedance with the far end short-circuited (Fig. 5-1).

Although this relationship is a measurement technique, rather than an ordinary operating mode, the impedance value thereby obtained for Z_0 has a very special significance in transmission-line theory. A line of any length that is terminated at its far end by Z_0 will not reflect energy back to the source (input), but will absorb all energy that reaches it. The only other situation where no reflections from the far end of the line back to the input occur is for an infinitely long transmission line. Thus, the Z_0 terminated line simulates the infinitely long line in this respect. In most RF applications, Z_0 is almost purely resistive, so it is often symbolized as R_0.

¼-λ lines

Transmission lines that are electrically ¼-λ display unique properties; these are readily put to good use in solid-state RF circuits. At the top of Fig. 5-2 are unbalanced and balanced versions of the ¼-λ impedance transformer. Other physical realizations of transmission lines, such as twisted wires or stripline elements, are equally applicable. The interpretation of the circuits is that a resistive termination of value R_1 can be transformed to value R_2 (or vice versa), provided that the characteristic impedance, Z_0, of the line is equal to $\sqrt{R_1 \times R_2}$. This relationship applies to both the unbalanced and balanced forms.

The center and bottom illustrations in Fig. 5-2 show how the ¼-λ lines can be used to simulate parallel and series-resonant tanks. These arrangements can be used as the frequency-determining circuit in oscillators, as RF chokes, or as bypass elements.

To produce a large transformation between the terminating resistances of the impedance transformer, it might be convenient to accomplish this in more than one

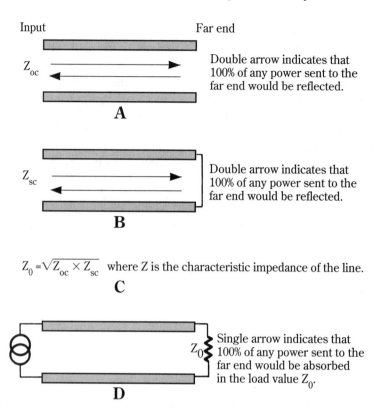

5-1 The characteristic impedance of a line from a classic measurement technique. A. Z_{oc} is input impedance with far end open-circuited. B. Z_{sc} is input impedance with far end short-circuited. C. Z_0 may be calculated from Z_{oc} and Z_{sc}. D. With Z_0 as far end termination (load), no power is reflected back to input.

step. Figure 5-3 shows cascaded ¼-λ lines used in this fashion. Resistance R_2 is a "phantom resistance" in the sense that it has no physical existence. Its value is, nonetheless, "seen" by both ¼-λ lines. Additional cascading can be used. Besides its use to provide large transformations, this technique is often used to develop broadbanded response.

In Fig. 5-3, the characteristic impedance (Z'_0) of the second line is higher than that of the first line (Z_0). This is symbolized by the wider spacing of the elements in the second line. Thus, it follows that R_2 is greater than R_1, and R_3 is greater than R_2. The arrangement transforms from a low resistance, R_1, to a high resistance, R_3, or vice versa. However, R_1 and R_3 cannot be interchanged.

From a mathematical viewpoint, it is not necessary to deal with the nonphysical resistance, R_2. The formulas for the two line impedances then become:

$$Z_0 = \sqrt{R_3 \times R_1^3} \quad \text{and} \quad Z_0' = \sqrt{R_1 \times R_3^3}$$

Broadbanding with ¼-λ impedance transformers A useful feature of the cascaded ¼-λ impedance transformer is its broadened frequency response. This broadbanding effect increases with the number of sections and is greater for lower overall

Characteristic impedance of transmission lines 157

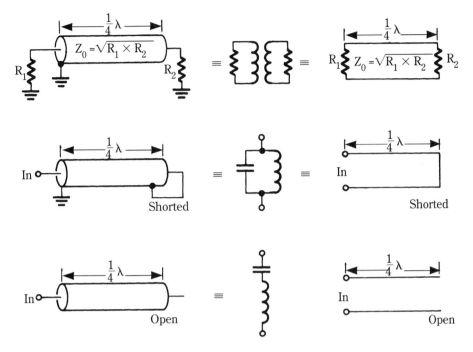

5-2 Unbalanced and balanced ¼-λ transmission line functions. Unbalanced circuits are on the left; balanced circuits are on the right. For the impedance transformer, the lines must have a certain characteristic impedance. For simulation of the *LC* resonant circuits, the value of the characteristic impedance is generally not critical.

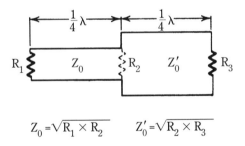

5-3 Cascaded arrangement of ¼-λ transformation lines. *R2*, a nonphysical resistance, is common to both ¼-λ sections. The overall transformation ratio is higher than readily attained with a single line.

ratios of impedance transformation. These characteristics are illustrated by the selectivity curves of Fig. 5-4. With a little ingenuity, you can implement simple, but effective, broadband impedance transformers at UHF and especially at microwave frequencies. The coiling of coaxial cable is permissible where problems are encountered with physical length. Microstrip transmission line does not necessarily need to be laid out as a straight run; it can incorporate bends to conserve linear space. Another useful aspect of the multiple-section impedance transformer is that its geo-

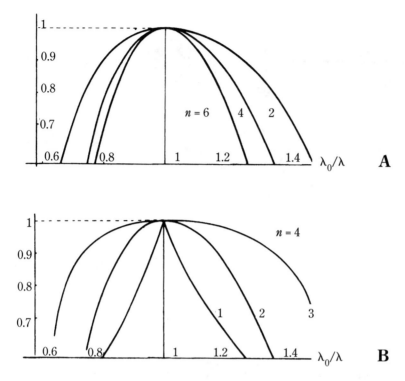

5-4 Selectivity curves for ¼-λ impedance transforming systems. Broadbanding increases with more sections and with lower overall ratios of impedance transformation. A. Two ¼-λ sections as shown in Fig. 5-3. B. One, two, and three ¼-λ sections; overall impedance transformation ratio, $n = 4$. Motorola Semiconductor Products, Inc.

metric and electrical parameters are less critical than single-section transformers in applications where broadbanding is not the first-priority operating feature.

¼-λ lines applied to push-pull circuits An almost-too-good-to-be-true use of ¼-λ lines is shown in Fig. 5-5. Here, the lines perform simultaneously as baluns, impedance transformers, and phasing circuits for proper push-pull operation of the two transistors. The arrangement is depicted in simplified form in order to focus attention on the implementation of the lines. Thus, various blocking capacitors, RF chokes, and reactance-canceling components are omitted. These, however, can be readily incorporated in the scheme in the same way as in more conventional amplifier circuits.

In this example, the resistive components of both the input and output impedance of the individual transistors are 6.25 Ω. Thus, both input and output lines "see" 12.5 Ω at their transistor terminations. The 25-Ω lines fit the situation beautifully, because $25 = \sqrt{12.5 \times 50}$. This takes care of the transformation feature of the lines.

The lines comply with the push-pull aspect of the amplifier circuitry. The base signals are 180° out of phase with one another. By reciprocal action, the 180° out-of-phase signals from the collectors are converged as a single composite signal in the output line.

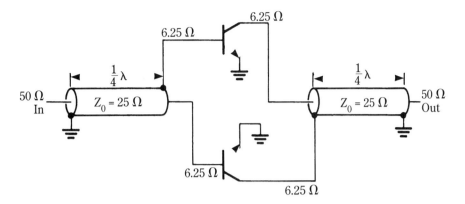

5-5 ¼-λ lines used for impedance matching, phasing, and as baluns in a push-pull amplifier. This system is exceptionally simple, but a fortuitous combination of transistor operating conditions and line impedances is needed.

Finally, these ¼-λ lines function as baluns. The two base signals are balanced, with respect to ground. The output signal delivered to the load is unbalanced to ground—even though the collector signals are balanced to ground.

¼-λ lines in transistor circuits The ¼-λ line provides its usually desired transformer action only when terminated in real (resistive) impedances. However, transistor input and output impedances are generally complex: they have both resistive and reactive components. To properly accommodate a ¼-λ line transformer, the reactive component must be tuned out. An example is shown in Fig. 5-6A, where a ¼-λ line is used to transform the resistive portion of the transistor input impedance to the desired 50 Ω. At the higher frequencies, the input reactance of transistors tends to be inductive from the effect of the internal bonding wires. Therefore, an equal and opposite reactance is incorporated to cancel the transistor's input reactance. This takes the physical form of a capacitor or a section of transmission line that is selected to simulate the required capacitance. Capacitor C serves this purpose in Fig. 5-6A. This makes the ¼-λ line perform properly, and it then provides transformation between the resistive component of the transistor's input impedance and the desired 50 Ω.

Figure 5-6B shows the same principle applied to the output of a transistor amplifier. Here, the undesired reactance is capacitive, so an inductive reactance is needed. This is often accomplished by a small lumped circuit inductor. From its schematic position, this inductor appears to function as an RF choke. However, instead of complying with the simple requisite of providing high inductive reactance, the stipulation here is for a discrete value of inductive reactance. Of course, such an inductor also can serve as a dc feed path. In practice, it is common to use both this resonating inductance and an RF choke, together with the usual decoupling circuitry. In any event, the ¼-λ line then performs its transformation between resistive terminations.

⅛-λ lines

Like the ¼-λ transmission line, one having an electrical length of ⅛ λ displays some uniquely useful circuit properties. Similarly, the use of these properties affords the designer and experimenter interesting ways to match impedances in solid-state cir-

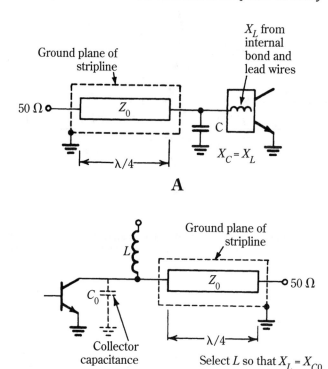

5-6 Use of ¼-λ lines in the input and output circuits of transistors. A. Application to input of UHF or microwave transistor. B. Application to output circuit of transistor. In both cases, λ/4 is the electrical length of the line and $Z_0 = \sqrt{R1 \times R2}$, where R1 and R2 are the terminating resistances "seen" by the line.

cuits. A peculiarity of the ¼λ is that its input end always "looks" resistive when its opposite end is terminated in the magnitude of the line's characteristic impedance. Notice that this statement allows complete latitude in how this impedance magnitude is composed; it can be made up of any combination of resistance and/or reactance that produces an impedance magnitude equal to the characteristic impedance of the line, Z_0. That is, the input to such a line will appear as a pure resistance if the terminated end "sees" the impedance

$$Z_0 = \sqrt{R^2 + X^2}$$

Here, Z_0 is the characteristic impedance of the line specified as a simple magnitude, such as 50 or 72 Ω. R and X are the termination resistance and reactance, respectively. The termination must be expressed as a series-equivalent circuit. If parallel-equivalent values are available for transistor impedances, they must first be converted to the series-equivalent format (see Table 4-1).

The fact that a resistive input can be achieved despite the presence of reactance at the terminated end is, indeed, interesting. But what about the value of the input resistance? It is found from the equation:

$$R_{IN} = \frac{R}{1 - \dfrac{X}{\sqrt{R^2 + X^2}}}$$

where R is the resistive component of the terminating impedance and X is the reactive component of the terminating impedance. Figure 5-7 shows an example in which the base of a transistor is the terminating impedance.

It is always possible to implement an ⅛-λ line to provide a purely resistive input. Microstrip lines enable considerable flexibility in this respect because they can be readily constructed to have characteristic impedances from several Ω to several hundred Ω. It is not always easy to produce the 50-Ω input impedance, however. This often involves appropriate selection of transistors and operating conditions. The ⅛-λ matching technique is applicable to the output of the transistor also. Notice in this example that had the terminating impedance been capacitive (that is, 20 - $j15$ Ω), the resistive input would have had a magnitude of 12.5 Ω, rather than 50 Ω (R_{IN} would then be 20/40/25 = 500/40 = 12.5).

⅛-λ lines as a reactance Another use of the eighth-wavelength line is as a substitute for a physical capacitor or inductor. Specifically, when an ⅛-λ is open-terminated, its input appears as a capacitance, with reactance numerically equal to the characteristic impedance, Z_0, of the line. Conversely, when the ⅛-λ line is short-circuited at its termination end, the input to the line appears as an inductance. Again, the inductive reactance is numerically equal to the characteristic impedance of the line.

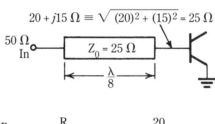

$$R_{in} = \frac{R}{1 - \dfrac{X}{\sqrt{R^2 + X^2}}} = \frac{20}{1 - \dfrac{15}{\sqrt{(20)^2 + (15)^2}}}$$

$$R_{in} = \frac{20}{1 - \dfrac{15}{25}} = \frac{20}{\dfrac{10}{25}} = \frac{500}{10}$$

$$R_{in} = 50 \ \Omega$$

5-7 A microstrip ⅛-λ line used to match base impedance to 50-Ω source resistance. Equal magnitudes of base impe**dance** and the characteristic impedance of the line result in a purely resistive input. The *magnitude* of this input resistance is determined by the reactive and resistive components of the base impedance. "Matching" is produced in this example because the base impedance is transformed to the desired 50-Ω input resistance.

A salient feature of such simulated reactance is its purity. Physical capacitors and inductors will inevitably display greater losses than their ⅛-λ line counterparts. Greater precision can be implemented in locating these line reactances in the network than is readily achieved with relatively bulky capacitors and inductors. Also, microstrip construction techniques provide a considerable range of reactance values; these are governed by width of the line element and the nature of the dielectric between it and the ground plane.

When such ⅛-λ lines are used in place of physical capacitors or inductors to cancel the input or output reactance of a transistor, they are known as *stub reactances*. Stub reactances are selected to present equal, but opposite, reactances to the parallel-equivalent reactance that is "seen" at the transistor terminal. Notice that this is different from the procedure used with ⅛-λ elements as impedance-matching components; there, the impedance at the transistor terminal was expressed in the series-equivalent format.

Figure 5-8 shows typical deployment of ⅛-λ lines as reactance stubs. In the input circuit, the ⅛-λ line is open at its "far" end and therefore acts as a capacitance. Conversely, the twin output stubs terminate in RF shorts, which are provided by dc-blocking capacitors. These stubs act as inductances. Notice that by feeding in the dc collector voltage at the RF-shorted end of one of these stubs, it is feasible to dispense with an RF choke in this portion of the circuit.

Additional attributes and uses of the ⅛-λ stubs Although an open-ended ⅛-λ stub simulates a capacitor at the operating frequency, its behavior differs markedly from the physical capacitor at the second harmonic. The reactance of the capacitor is merely one half of the value that pertains to the fundamental frequency. The capacitive stub line, however, appears as a short to the second harmonic. This can be seen by noting that such an ⅛-λ line becomes a ¼-λ line at the second-harmonic frequency. The nature of a ¼-λ line with its far end open to present a short circuit to the source. Because impedance-matching networks are also exploited for their har-

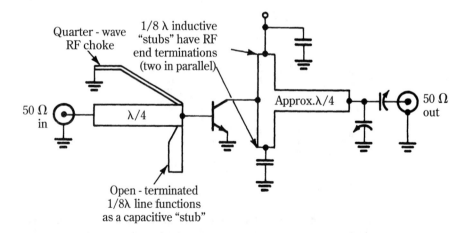

5-8 The typical use of ⅛-λ stubs for simulating capacitance or inductance. The transistor is construed to require capacitive reactance to cancel its inductive input reactance and to require inductive reactance to cancel its capacitive output reactance.

monic-attenuating capabilities, such behavior of the ⅛-λ capacitive stub is destined to have practical implications. One of these is that superior attenuation of second-harmonic energy is obtained over networks that use physical capacitors in their shunt arms. Therefore, the overall filtering problem is often relaxed.

It has also been found that the ⅛-λ capacitive stub can be used in an output network to significantly improve the collector efficiency of the transistor. For this to be accomplished, the first element in the output network must be a series inductance. Networks such as those shown in *A*, C_2, and *D* of Table 4-2 can be used or readily modified to attain this objective. The basic idea is that the first inductive element isolates the physical collector from the first capacitive shunt element. This results in a buildup of harmonics so that the collector voltage approaches a square-wave shape. The transistor then functions more like an ideal switching device, and dissipation from the simultaneous presence of both collector voltage and collector current is reduced. At low frequencies, this enhanced collector efficiency can be realized with L and tee networks that use ordinary capacitors in the shunt arms. At VHF, UHF, and microwave frequencies, the ⅛-λ stub produces better results.

Figure 5-9 shows how the ⅛-λ stub is incorporated for this purpose. Its physical location along the series inductive or impedance-transforming element is best determined empirically. The ideal position of this tap appears to be on the order of ¹⁄₁₆ λ from the collector of the transistor. With this technique, collector efficiencies exceeding 80% have been achieved from 25-W amplifiers operating at 400 MHz.

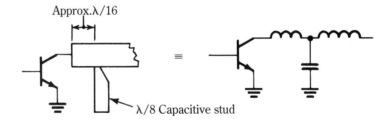

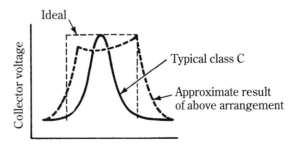

5-9 The technique for obtaining increased collector efficiency. By providing a small amount of inductance at collector, the harmonic composition of the collector voltage can be tailored to develop a closer approach to a rectangular wave.

Other lines for impedance matching

The use of ¼-λ and ⅛-λ lines as impedance transformers has been described. Other transmission-line lengths can often be applied for this function. Again, however, certain criteria must be met.

Transmission-line elements somewhat shorter than ¼ λ display the useful property that direct transformation can be accomplished from a purely resistive termination to one that contains inductive reactance. Thus, such a line can often be used between a 50-Ω input source and the base of the driven power transistor. The mathematics of such an impedance match indicates that no stubs or other reactances are needed. In order for the match to be realized, two equations must be satisfied:

$$Z_0 = \sqrt{R1 \times R2} \times \sqrt{1 - \frac{(X2)^2}{R2(R1-R2)}}$$

and

$$\tan \beta 1 = Z_0 \left(\frac{R1 - R2}{R1 \times R2} \right)$$

where:
Z_0 = characteristic impedance of the line.
R_1 = resistive termination of the line (usually 50 Ω).
R_2 = resistive component of the transistor impedance.
X_2 = reactive component of the transistor impedance (must be inductive).
β_1 = electrical length of the line in degrees

Some insight into these equations can be attained by supposing that we have a situation using a ¼-λ line, which provides transformation between R1 and R2. In such a case, X_2 as well as X_1 is necessarily zero. This reduces the first equation to the simpler form: $Z_0 = \sqrt{R1 \times R2}$, which is correct for ¼-λ lines used as impedance transformers. Because X_2 is zero, the line length corresponds to the tangent of infinity, which is 90 electrical degrees, or a ¼-λ length. Thus, these equations are generalized relationships that embrace the ¼-λ situation.

Further consideration reveals another aspect of these equations. It can be seen that X_2 cannot be too large, for otherwise the term $(X_2)^2/[R_2(R_1 - R_2)]$ will exceed unity and Z_0 will not be realizable. It can also be appreciated that the reactance represented by X_2 must be inductive, not capacitive. Otherwise, the calculation for line length cannot be made. Despite these limitations, the use of such a "shortened" ¼-λ line is very desirable where applicable.

Figure 5-10 depicts a typical situation in which a line section shorter than an electrical ¼-λ length accomplishes impedance transformation between a purely resistive 50-Ω source and the complex impedance of the base of a transistor. In a qualitative way, you can postulate that the inductive component of the transistor input impedance replaces a portion of what would otherwise be a ¼-λ line. The 0.1774 wavelength pertains to electrical length or to the length under ideal conditions in air. Because stripline construction is implied, the effective dielectric constant of the insulation material must be taken into account when determining the physical

Characteristic impedance of transmission lines 165

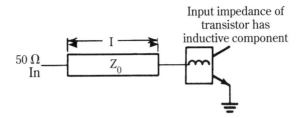

Example: Calculate line to transform input impedance of transistor = 7 + j8 Ω to 50 Ω.

$$Z_0 = \sqrt{R_1 \times R_2} \times \sqrt{1 - \frac{(X_2)^2}{R_2(R_1 - R_2)}}$$

$$= \sqrt{50 \times 7} \times \sqrt{1 - \frac{8^2}{7(50 - 7)}}$$

$$= \sqrt{350} \times \sqrt{\frac{64}{7(43)}} = 18.71 \times \sqrt{1 - 0.213}$$

$$= 18.71 \times \sqrt{0.787} = 18.71 \times 0.887$$

$$= 16.6 \text{ Ω, a value readily attainable with stripline techniques}$$

Then, $\tan \beta I = Z_0 \left(\frac{R_1 - R_2}{R_1 \times R_2} \right) = 16.6 \left(\frac{50 - 7}{50 \times 7} \right)$

$$\tan \beta I = 16.6(43/350) = 16.6(0.1229)$$

$$\tan \beta I = 2.0394, \beta I = \text{Arctan } 2.0394 = 63.883°$$

and:

$$I = 63.883\lambda/360 = 0.1774\lambda$$

5-10 Direct impedance transformation with a "shortened" ¼-λ line. Under appropriate conditions, the inductive component of the transistor input impedance can replace a portion of a line which otherwise would have to be ¼-λ of an electrical wavelength.

length of the line. If, for example, the effective dielectric constant is 2.5, the physical length of the line will be very nearly

$$\frac{0.1774}{\sqrt{2.5}} = 0.112 \text{ wavelength,}$$

as measured in air. The electrical wavelength remains 0.1774, of course.

In principle, the inductive reactance of the transistor used in this example could have been as high as approximately 17.3 Ω, but awkward design parameters would accrue before this value was closely approached. Even less design latitude is available with the use of coaxial lines because of the relatively limited characteristic impedance values that are available. It is permissible to connect coaxial lines in parallel, however. Another way of gaining design flexibility is to use twisted-pair lines, where the characteristic impedance can be manipulated by the tightness of the twist. Electrical losses and mechanical problems tend to occur in twisted lines, however.

½-λ **lines** A transmission line with an electrical length of ½ λ behaves as an impedance repeater (Fig. 5-11). In Fig. 5-11A, the input impedance of the ½-λ line is Z_2, the load impedance at the far-end termination. Conversely, the output impedance of the ½-λ line in Fig. 5-11B is Z_1, the source impedance. Interestingly, the characteristic impedance, Z_0, of the line does not affect this behavior. Moreover, Z_1 and Z_2 does not need to be resistive, but can have any impedance value, including 0 (a short-circuit) or infinity (an open-circuit). Notice that the situations depicted in Fig. 5-11A and B are separate and independent operating modes; one arrangement does not bear on the other.

This "transparency" of the ½-λ line exists also for longer lines, which are integral multiples of a ½ λ. Thus, a one-λ line and ¾-λ lines will repeat impedances in the same fashion. If the line is many ½ wavelengths, as in a long antenna feedline, the additional factor of attenuation might have to be taken into account in the delivery of power from the source to the load.

Yet another way of looking at the short-circuited ½-λ transmission line is that it can simulate the action of a series-resonant circuit. In such a circuit, the source "sees" a very low impedance, ideally zero, at the resonant frequency. Conversely, the open-circuited ½-λ line can simulate the action of a parallel-resonant circuit, in which case the source works into a very high impedance (ideally infinite) at the resonant frequency. Because these resonant circuit simulations can be accomplished with ¼-λ lines, ½-λ lines are not commonly found in tank circuits of amplifiers or oscillators.

A little thought will show that there is no ambiguity between ½- and ¼-λ line be-

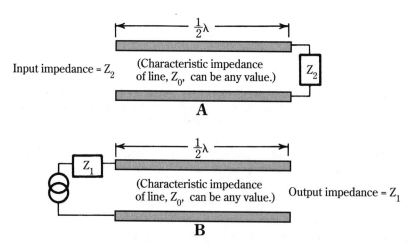

5-11 Impedance-repeating feature of the ½-λ transmission line. A. Impedance "seen" at input of ½-λ line is far-end or load impedance, Z_2. B. Impedance "seen" at the far end of ½-λ line is the input impedance, Z_1.

havior. ¼-λ behavior manifests itself only for odd multiples of a ¼-λ. Thus, you can obtain similar ¼-λ behavior from lines that are ¼, ¾, ⁵⁄₄, or any odd number of ¼-λ lengths. None of these lines can be said to be also ½-λ lines. For example, a line that is three ½ λ long is also accurately described as a six ¼-λ line, but as such is not an odd multiple of ¼ wavelengths. Accordingly, such a line would not display ¼-λ behavior, but would behave essentially as a simple ½-λ line. At the same time, it can be appreciated that any line can be made to display either ¼- or ½-λ characteristics if the frequency from the driving source is varied sufficiently.

Transmission-line connections between driver and power amplifiers

Figure 5-12 illustrates commonly encountered operating conditions when a transmission line is used between the driver and power amplifier. The flat line of part A is the best because a condition of impedance match prevails regardless of the length of the line or the frequency. Situation B also results in a flat line, but maximum available power cannot be obtained from the driver. The VSWR is unity for situations A and B.

In Fig. 5-12C, standing waves will be on the line because of reflections from the mismatched load impedance. A line length can be found wherein maximum power can be delivered to the load under these conditions. The energy transport efficiency of the system will, at best, be lower than that attainable in situations A and B. Once the line length is optimized, it will not hold if the frequency is appreciably changed. Thus, for broadband work, the situations depicted in A or B are very desirable.

Figure 5-12D also gives rise to standing waves. Here, again, optimized energy transport is attainable at a discrete line length. In this case, however, the maximum available power can be extracted from the driver when the line length is optimum. In this respect, it is as efficient as is the operating mode of situation A. However, the optimum operating condition is, unlike situation A, sensitive to both line length and frequency.

Figure 5-13 shows a special case for the generalized operating mode of Figure 5-12C. In this case, the transmission line behaves as a ¼-λ transformer, and an overall impedance match is established. Because of the transformer action, the maximum available driver power is available for delivery to the input of the power amplifier. The efficiency is essentially as good as in Figure 5-12A, but, of course, with sensitivity to line length and frequency. As an example, suppose that 50-Ω coaxial cable is used. Then, a close impedance match can be attained if the driver output impedance is 60 Ω resistive and the power amplifier input impedance is about 41.7 Ω. That is, $Z_0 = 50 = \sqrt{60 \times 41.7}$. Many other pairs of impedance values can satisfy the basic requirement, which is $\sqrt{Z_1 \times Z_2} = Z_0$. Thus, $\sqrt{90 \times 27.78} = 50$ Ω. Notice that you have the option of going from a 90-Ω source to a 27.78-Ω load, or from a 27.78-Ω source to a 90-Ω load. The ¼-λ transmission-line transformer is equally adaptable, whether stepping up or down.

Flat lines A flat transmission line is one in which no standing waves exist, that is, one operated so that the VSWR is unity. This operating condition occurs when the source impedance, the characteristic impedance of the line, and the load impedance are all the same value. Moreover, because the characteristic impedance is essentially resistive, the source and load impedances must be resistive, too. In practice, this gen-

168 *Applications of transmission-line elements to RF power circuitry*

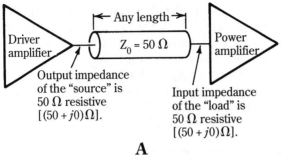

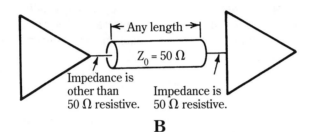

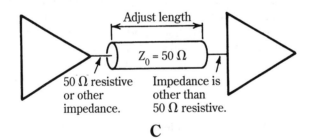

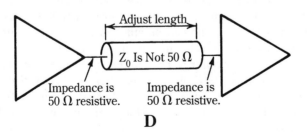

5-12 Commonly encountered line situations in driver-power amplifier connections. A. Ideal situation: flat line with maximum energy transport. B. Line is flat, but less than maximum drive power is available. C. Standing waves on line: line is sensitive to both length and frequency. D. Standing waves on line: the line is sensitive to both length and frequency.

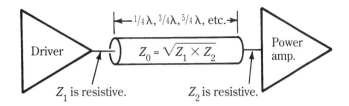

5-13 A special case of situation C in Fig. 5-12. Here, the transmission line functions as a ¼-λ transformer, providing an impedance match for source and load. Energy transport is maximized, but the match is sensitive to line length and frequency.

erally involves 50-Ω coaxial cable and 50-Ω resistive source and load. Ideally, such a line constitutes the most efficient and the least trouble-prone method of feeding energy from a driver to an output amplifier or from the output amplifier to the load. One of the most useful features of the flat line is that its conveyance of energy is not frequency sensitive. An equally important corollary of this feature is that the operation of such a line is not a function of its physical or electrical length. That is, making the flat line longer or shorter will not disturb the unity VSWR. Indeed, unity VSWR is the operating condition that defines the flat line.

The "flatness" of a transmission line has nothing to do with its balance, with respect to ground. Thus, both two-wire lines and coaxial-cable lines might be operated in their flat modes, even though the former might be balanced and the latter, unbalanced.

Tuned lines Energy might still be transported via transmission lines, which are not subject to the ideal terminating conditions that were described previously. Generally, this comes about because of wrong impedance at the load. Counterreactance can then be inserted in such a way that the transmission line becomes part of a resonant system. Because of the high SWR, which might attend such operation, electrical losses in the line tend to be greater than in the flat line. If the line is not too long and if its dissipative losses are low, it might still operate quite efficiently in this mode. However, such a line is inherently frequency- and length-sensitive. The efficiency of a system using a tuned line will be at its maximum attainable value when the resistive component of the source impedance, the characteristic impedance, and the resistive component of the load impedance are nearly equal (low VSWR). Less efficiency will result if only the resistive components of source and load are close. If even this is not the case, the efficiency of energy transport from source to load will be further impaired. The line can still be resonated, however, for optimal efficiency under these conditions.

The primary application of the tuned line is as feeder between the output amplifier and the antenna. However, with the inordinate emphasis on low VSWR, much effort is expended in making these feedlines operate as flat as possible. They are "tuned" only in the sense that residual reactance is neutralized by the adjustment provisions. Tuned lines with high VSWR are, nonetheless, satisfactory if their higher losses, as well as frequency and length sensitivity, can be tolerated.

Transmission-line transformers The *transmission-line transformer* is a unique and useful RF component. Its physical resemblance to a conventional transformer is deceiving, because its operating principle is quite different. Its performance, particularly its efficiency and broadband response, clearly indicate that more than meets the eye is involved. A common physical form consists of a few turns of paired wires on a toroidal core. Casual inspection would suggest it to be an ordinary bifilar wound transformer. The winding connections might be suggestive of an autotransformer.

Closer involvement with such a transformer could turn up some strange features. The magnetic core is nearly passive as a means of energy transfer between the input and output terminals. Mysteriously, the degradation in high-frequency response from interwinding capacitance is negligible. The number of turns do not stem from classical transformer design equations. If you have had considerable experience with ordinary transformers, you might be surprised at the high power levels that a compact device of this nature can handle, all the while running cool. Finally, there are very low level nonlinear effects—even when operating at high power.

The preceding facts, and more, are attributable to the fact that the "windings" do not function as inductively coupled coils, but rather as short transmission lines; generally at the highest frequency of operation. These lines are still less than an ⅛-λ length. The coiled or helical format of these lines has no direct electrical significance because a transmission line can be coiled without significantly altering its electrical characteristics. This is especially true for coaxial cable, but for practical purposes, it is substantially true of other line formats (such as twisted wire, or closely paired wire). Coiling simply compacts the physical dimensions of the lines. In so doing, however, it indirectly contributes to the electrical realization of the device; it permits certain terminals to be connected together almost directly.

Figures 5-14 and 5-15 show some common versions of transmission-line transformers. In all of these, the characteristic impedance of the line, regardless of its type, is the geometric mean of the input and output resistances; that is, $Z_0 = \sqrt{R_{IN} \times R_{OUT}}$. Because these are all 4:1 or 1:4 impedance-transforming devices, by substitution:

$$Z_0 = R \times \frac{R}{4} = \sqrt{\frac{R^2}{4}} = \frac{R}{2}$$

Thus, Z_0 is selected or fabricated to be one-half the value of the larger terminating resistance. Or, stated another way, Z_0 is twice the value of the smaller terminating resistance. Here is a difference in the application of these devices from that of conventional transformers, where considerable latitude pertains to the terminating impedances. In actual practice, the three impedances involved (R_{IN}, R_{OUT}, and Z_0) in the application of the transmission-line transformer often deviate somewhat from the mathematical ideal. Surprisingly, the bandwidth and the transformation ratio will often remain within acceptable limits.

Both cored and coreless devices are shown in the several illustrations. Do not suppose that the cores "load" the transmission lines because neither the characteristic impedance nor the electrical wavelength are affected by the magnetic cores. Moreover, those cores do not participate in the transfer of energy as in a conventional transformer. The core material is not inserted between the line elements, but rather it is adjacent to them. The use of a core does not significantly alter performance at higher frequencies, but it appreciably extends the low-frequency operating ability.

Figure 5-16 is the usual schematic diagram that is used to represent the trans-

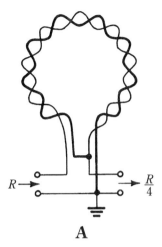

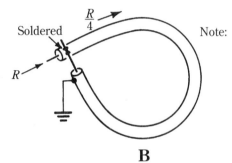

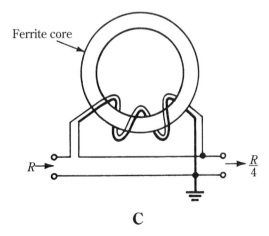

5-14 Three simple implementations of transmission-line transformers. A. Twisted wire. B. Coaxial cable. C. Paired wire on a ferrite core. Twisted wire or coaxial cable can also be used in this fashion.)

172 *Applications of transmission-line elements to RF power circuitry*

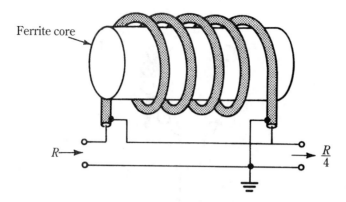

5-15 A transmission-line transformer wound on cylindrical core. Although coaxial cable is shown, this format can be implemented with paired or twisted wires as well.

mission-line transformers of Figs. 5-14 and 5-15. The coil symbols should not be construed to imply inductive coupling between the transmission-line elements—as between the primary and secondary of conventional transformers. However, this symbology is useful in depicting that the line elements, in addition to behaving as a transmission line, also exhibit ordinary inductance from end to end. This is important because one line element is shunted directly across either the source or the load. How can this situation prevail without having a short circuit?

The answer is that a line element, such as the one symbolized by the "inductor" 3-4 of Fig. 5-16, does indeed present ordinary inductive reactance to the flow of such shunt current. The essential thing to grasp here is that this inductive reactance stems from this line element's existence as a conductor. It has nothing to do with this line element's participation with line element 1-2 to form a transmission line. This being the case, it follows that the low-frequency shunting action of line element 34 would be reduced if its inductance could be increased. Then, you should expect greater low-frequency response from the transmission-line transformer. But how can this be brought about without changing the characteristics of the transmission line itself? One way is to coil the transmission line. This expedient, however, is generally necessary anyway so that the terminal connecting wires are short. If, furthermore, the

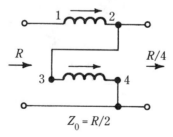

5-16 Schematic diagram of the transmission-line transformers of Figs. 5-14 and 5-15. Such a representation is useful to depict the terminal connections, but should not be construed to imply the kind of inductive coupling associated with ordinary transformers.

line is coiled around ferrite or other suitable magnetic material, the inductance of the single elements will be greatly increased, but there will be virtually no effect on the transmission line parameters. As pointed out previously, the ferrite would have to be inserted between the line elements in order to affect the characteristic impedance or the electrical line length.

From what has been described, the desired effect of the ferrite core is to make such a single-line element as 3-4 a more effective "RF choke" to low-frequency shunt currents. Although early literature suggested transmission-line action at high frequencies, and ordinary transformer action at low frequencies, present thinking tends to ascribe operation in the transmission-line mode throughout the passband of the device. The core merely extends the low-frequency response via the described mechanism. Figure 5-17 shows the general effect of a core. In practice, there is much less tendency for temperature rise in the core, compared to the situation with an ordinary transformer operating under similar power conditions. This reinforces the theory that the core does not participate significantly in the transfer of energy from input to output. A corollary of this is that the core does not need to be made of exotic high-frequency material, and it can be of smaller volume than in a conventional transformer with similar ratings.

1:1 transmission-line transformers The arrangement shown in Fig. 5-18 does not transform impedance values, but it exhibits the interesting property of functioning as a transposition device between balanced and unbalanced circuitry. That is, it is inherently a *balun*. At first inspection, it would appear that this arrangement differs from the 4:1 configuration of Fig. 5-16 only in the connection of its leads. Qualitatively, this is so. However, the characteristic impedance in this case is R, rather than $R/2$ Ω. In both of these transmission-line implementations, Z_0 is the geometric mean of the input and output impedances. Thus, if the input or output impedance changes because of a different transformation ratio, it follows that a different Z_0 must be used.

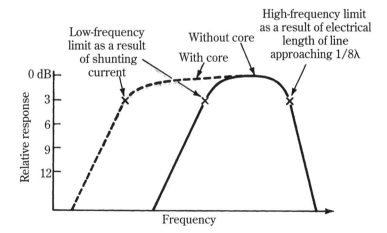

5-17   The general effect of adding a magnetic core to a transmission-line transformer. Low-frequency response is extended because more inductive reactance is provided to impede the flow of shunting currents.

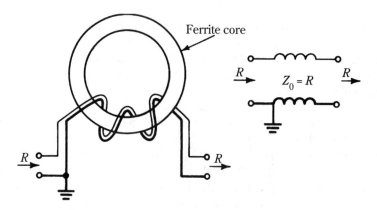

5-18 A 1:1 transmission-line transformer and its schematic diagram. The primary use of this arrangement is to convert from unbalanced to balanced circuitry, or vice versa.

A definite resemblance exists between this transmission-line balun and the bifilar-wound RF choke used in the filament circuit of grounded-grid tube amplifiers. The operation of the latter component is predicated on its high inductance, and any "characteristic impedance" mutual to the windings plays no direct part in providing high impedance to RF. Moreover, such a choke is used as a shunt element, with respect to the RF. Conversely, the 1:1 transmission-line transformer must have a discrete characteristic impedance, and be used as a series element to transport RF with minimal, rather than with high, attenuation. Although its inductance is important in broadbanding its performance, inductance is not the basic parameter underlying its operation, as it is in the bifilar-wound RF choke.

Transmission-line transformers for other impedance ratios If multiple line elements are properly interconnected, other impedance transformation ratios than 4:1 or 1:4 can be realized. Figure 5-19 illustrates the technique for devising transmission-line transformers with impedance ratios of 9:1 (or 1:9) and 16:1 (or 1:16). Although the general scheme can be extended to ratios of 25 and 36, practical difficulties of construction tend to make the higher ratios somewhat difficult to implement.

In its simplest format, these transmission-line transformers consist of twisted wires: three for the 9:1 transformer and four for the 16:1 transformer. Despite the multiplicity of turns in the schematic diagrams, these transformers often use a single turn, or slightly less than a single turn of the twisted wires. This stems from the rule that the length of the wires should be no more than about one $\frac{1}{8}\lambda$ at the highest frequency of interest. If desired, a ferrite core can be used to extend the transformer's low-frequency operation. Generally, some empirical work must be done to obtain optimum results, particularly if broadbanding is the objective. For example, some experimentation is usually needed to determine the way in which the wires are twisted together, because this directly governs the characteristic impedance of adjacent wires. Fortunately, the parameters are not critical for most applications, and it is usually easy to attain better results than with a conventional electromagnetic-type transformer.

As would be expected, the phasing of the line elements is all-important. The enumeration shown in the diagrams of Fig. 5-19 corresponds to the ends of the twisted wires. Even numbers correspond to one end; odd numbers correspond to the opposite end.

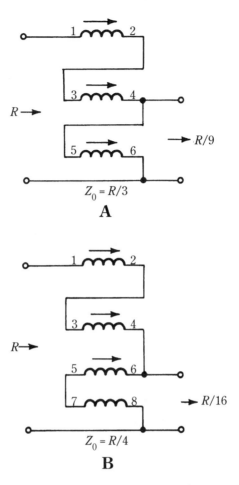

5-19 Diagrams of 9:1 (or 1:9) and 16:1 (or 1:16) transmission-line transformers. A. Arrangement for impedance transformation by factor of 9 or ⅑. B. Arrangement for impedance transformation by factor of 16 or ¹⁄₁₆.

Various constructions can actually be used. Instead of twisted wires, parallel wires or multifilar winding formats are applicable. Either a solenoidal or toroidal core might also merit consideration. As with the 4:1 line transformer (previously described), the basic operating principle is premised on the conductors behaving as short transmission lines. Accordingly, it can be misleading to think of the "windings" as being inductively coupled in the manner of conventional electromagnetic transformers. That this is not so is readily inferred by the relatively cool operation of a core, if used, even when handling high power levels. Eddy current and hysteresis loss in the core are minimal because, unlike in the conventional transformer, the core does not participate in the energy transfer between the "windings."

Special considerations when designing with stripline elements

The simple formulas found in handbooks for calculating the characteristic impedance of two-wire and coaxial transmission lines are not generally applicable to the stripline structure. The primary reason is that this structure gives rise to complex fringing patterns of the electric lines of force and to proximity effects. This, in turn, results in an effective value of the relative dielectric constant that is lower than would be measured by incorporating the board material in a capacitor "sandwich" in which both copper plates have extensive and equal areas. In the stripline structure (Fig. 5-20), ratio $W{:}H$ exerts a strong effect on the effective value of the relative dielectric constant. In all transmission lines, the characteristic impedance is governed by both geometric factors and by the relative dielectric constant. Because the geometry of the line directly influences the characteristic impedance via physical dimensions, and also by its effect on the effective value of the relative dielectric constant, the mathematical format connecting Z_0 with geometrical dimensions will be more complex than simple.

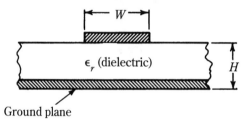

5-20 The basic physical and electrical parameters in stripline (or microstrip) circuit elements. Before determining the length of the line, these parameters must be taken into account because they influence the characteristic impedance and the velocity factor.

Also, in all transmission lines, the velocity factor is a function of the relative dielectric constant. For most practical purposes, the velocity factor in air is very close to 1.00. The insertion of dielectric material other than air decreases both the characteristic impedance and the velocity factor. The practical effect of reduced velocity factor is that, for a given electrical length, the physical length of the line will not be the same as it would be with an air dielectric. Specifically, the physical length will be reduced by the factor:

$$\frac{1}{\sqrt{\epsilon_r}}$$

where ϵ_T is the effective value of the relative dielectric constant. This, in most cases, is a favorable situation because it enables more compact construction. For example, a ¼-λ line that would require, for example, 10 cm of air-dielectric transmission line (dielectric constant = 1.00) would require a length of only 5 cm if it was constructed with a dielectric material that had a relative dielectric constant of 4. Also, the new velocity factor would be 0.5, rather than unity.

Figure 5-21 depicts characteristic impedance, Z_0, as a function of the ratio $W{:}H$ for striplines that use several different dielectrics. Figures 5-21, 5-22, and 5-23 show the variation in the effective value of the relative dielectric constant. For high $W{:}H$ ratios, the effective value of the relative dielectric constant approaches that of the material itself. Similar information is given in tabular form in Table 5-1, in which the velocity factors are also indicated. The use of such curves and tables greatly simplifies the design of stripline elements.

Table 5-1. Microstrip Z_0 and velocity factor versus width-to-height (W/H) ratio[a]

W/H	Air $\varepsilon_R = 1.0$		Teflon $\varepsilon_R = 2.55$		Epoxy $\varepsilon_R = 4.25$		Alumina $\varepsilon_R = 9.6$	
	Z_0	V_P	Z_0	V_P	Z_0	V_P	Z_0	V_P
0.630	168.425	1.000	110.683	0.657	87.986	0.522	60.977	0.362
0.695	161.878	1.000	106.258	0.656	84.414	0.521	58.441	0.361
0.766	155.370	1.000	101.865	0.656	80.870	0.521	55.927	0.360
0.844	148.909	1.000	97.509	0.655	77.360	0.520	53.440	0.359
0.931	142.506	1.000	93.199	0.654	73.888	0.518	50.985	0.358
1.026	136.171	1.000	88.941	0.653	70.463	0.517	48.566	0.357
1.131	129.916	1.000	84.745	0.652	67.090	0.516	46.187	0.356
1.247	123.753	1.000	80.616	0.651	63.775	0.515	43.853	0.354
1.375	117.692	1.000	76.565	0.651	60.524	0.514	41.568	0.353
1.516	111.746	1.000	72.597	0.650	57.345	0.513	39.337	0.352
1.672	105.926	1.000	68.721	0.649	54.243	0.512	37.164	0.351
1.843	100.242	1.000	64.944	0.648	51.223	0.511	35.053	0.350
2.032	94.706	1.000	61.273	0.647	48.291	0.510	33.007	0.349
2.240	89.327	1.000	57.714	0.646	45.451	0.509	31.030	0.347
2.470	84.115	1.000	54.271	0.645	42.709	0.508	29.123	0.346
2.723	79.076	1.000	50.951	0.644	40.066	0.507	27.289	0.345
3.002	74.218	1.000	47.757	0.643	37.527	0.506	25.531	0.344
3.310	69.546	1.000	44.692	0.643	35.094	0.505	23.849	0.343
3.649	65.065	1.000	41.759	0.642	32.768	0.504	22.244	0.342
4.023	60.779	1.000	38.959	0.641	30.550	0.503	20.716	0.341
4.435	56.689	1.000	36.292	0.640	28.440	0.502	19.266	0.340
4.890	52.796	1.000	33.760	0.639	26.439	0.501	17.892	0.339
5.391	49.100	1.000	31.360	0.639	24.544	0.500	16.594	0.338
5.944	45.600	1.000	29.091	0.638	22.755	0.499	15.370	0.337
6.553	42.291	1.000	26.952	0.637	21.069	0.498	14.218	0.336
7.224	39.173	1.000	24.938	0.637	19.485	0.497	13.138	0.335
7.965	36.233	1.000	23.047	0.636	17.998	0.497	12.125	0.335
8.781	33.484	1.000	21.275	0.635	16.606	0.496	11.179	0.334
9.681	30.904	1.000	19.618	0.635	15.305	0.495	10.295	0.333
10.674	28.491	1.000	18.071	0.634	14.091	0.495	9.472	0.332
11.768	26.240	1.000	16.629	0.634	12.961	0.494	8.707	0.332
12.974	24.143	1.000	15.288	0.633	11.911	0.493	7.996	0.331
14.304	22.192	1.000	14.043	0.633	10.937	0.493	7.338	0.331
15.770	20.381	1.000	12.888	0.632	10.033	0.492	6.728	0.330
17.387	18.702	1.000	11.818	0.632	9.198	0.492	6.164	0.330
19.169	17.148	1.000	10.830	0.632	8.425	0.491	5.644	0.329
21.133	15.172	1.000	9.917	0.631	7.713	0.491	5.164	0.329
23.300	14.385	1.000	9.074	0.631	7.056	0.490	4.722	0.328
25.688	13.162	1.000	8.299	0.630	6.451	0.490	4.315	0.328
28.321	12.036	1.000	7.585	0.630	5.894	0.490	3.942	0.327
31.224	10.999	1.000	6.929	0.630	5.383	0.489	3.598	0.327
34.424	10.047	1.000	6.326	0.630	4.914	0.489	3.284	0.327
37.953	9.172	1.000	5.773	0.629	4.483	0.489	2.995	0.327
41.843	8.370	1.000	5.266	0.629	4.089	0.489	2.731	0.326
46.132	7.634	1.000	4.801	0.629	3.727	0.488	2.489	0.326
50.860	6.960	1.000	4.376	0.629	3.397	0.488	2.267	0.326

Table 5-1. Continued

56.073	6.343	1.000	3.987	0.629	3.094	0.488	2.065	0.326
61.821	5.779	1.000	3.632	0.628	2.818	0.488	1.880	0.325
68.157	5.264	1.000	3.307	0.628	2.566	0.487	1.711	0.325
75.144	4.792	1.000	3.010	0.628	2.335	0.487	1.557	0.325
82.846	4.362	1.000	2.739	0.628	2.125	0.487	1.417	0.325
91.337	3.969	1.000	2.492	0.628	1.933	0.487	1.289	0.325
100.700	3.611	1.000	2.267	0.628	1.758	0.487	1.172	0.324

Reflectometer-type instruments

Having investigated the unique nature of the transmission-line transformer, a few words about reflectometer instruments are apropos; these compose the so-called "reflectometer measuring devices," such as VSWR indicators, reflection-coefficient meters, return-loss meters, and directional wattmeters. All of these are predicated on some form of directional coupler—probe elements or circuit components that are capable of selectively distinguishing between forward and reflected traveling waves on transmission lines. This is a very useful behavior because the all-important standing waves on a transmission line are actually the result of the interference effects of the forward and reflected traveling waves on the line. A flat line (one that has a matched load) produces no reflected wave; therefore there is no interference with the forward-going RF energy and there is no standing-wave pattern. The voltage SWR of such a transmission line is unity because of the mathematical definition of VSWRs:

$$\text{VSWR} = \frac{V^+ + V^-}{V^+ - V^-}$$

where: V^+ is the forward voltage on the line.
V^- is the reflected voltage on the line.
VSWR can vary between 1.0 and infinity.

Thus, appropriate detector and metering circuits that are associated with directional couplers can provide requisite data for determining VSWR on a line. With such information, it becomes possible to apply tuning and transforming techniques to enable the RF source and its transmission line to "see" an impedance-matched load. Such a load or load system will then absorb maximum available power from the source. Also, transmission-line losses will then be minimal.

Interestingly, the concept and implementation of directional selectivity to electromagnetic waves has had a long history. In particular, an early example was the Beverage antenna. In its simplest form, this antenna is a single wire that is a few wavelengths long and elevated a few feet above ground level (Fig. 5-24). This is an open-wire transmission-line with the earth substituting for the missing conductor. This simulated transmission line is terminated by a resistance that is equal to the line's characteristic impedance at the transmitter end, and is fed to the receiver at the opposite end. Oncoming waves produce a signal at the receiver; waves received from

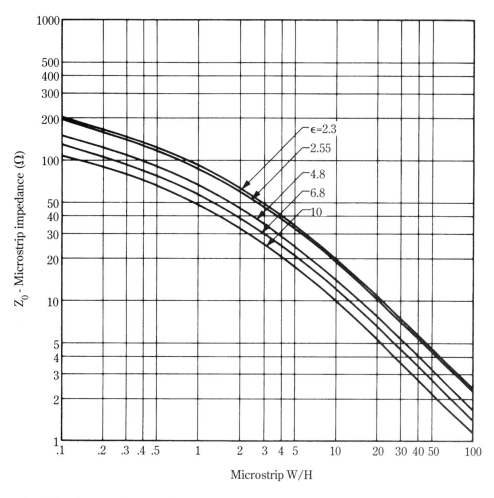

5-21 The characteristic impedance of microstrip transmission line versus W/H. By appropriately selecting the dielectric material and by design of the geometric features of the line, a wide range of Z_0 values can be realized. <small>Communications Transistor Corp.</small>

the opposite direction dissipate their energy in the terminating resistance. Waves that cross the line from the sides produce a small effect. Notice how different such behavior is from "ordinary" single-wire antennas. The Beverage antenna is easiest to implement at frequencies below several hundred kHz.

Monomatch bridge circuits A much-used reflectometer bridge for VSWR and RF power-indicating instruments is the circuit shown in Fig. 5-25. The two pick-up elements function as directional couplers, and are somewhat similar to Beverage antennas. However, whereas the Beverage antenna is generally a few to many wavelengths long, the pick-up elements in this measuring instrument, known as the *monomatch bridge*, are a small-fraction of a wavelength long. This is sufficient for the 180° directivity needed. As you might suspect, the sensitivity of the instrument is fre-

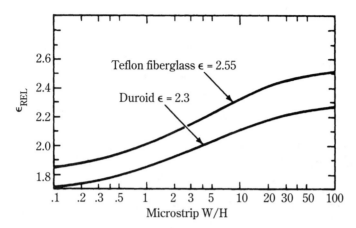

5-22 The effective value of dielectric constant versus W/H for low dielectric constant materials. For high values of W/H, the effective dielectric constant approaches the "rated" dielectric constant of the material. _{Communications Transistor Corp.}

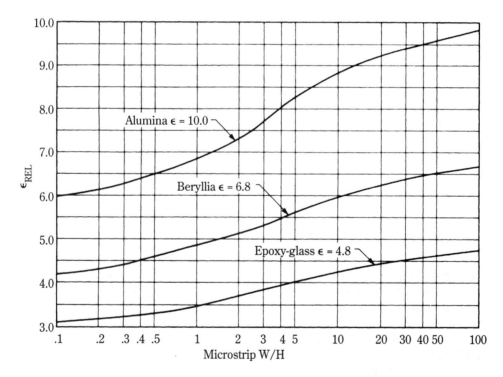

5-23 Effective value of dielectric constant versus W/H for high dielectric constant materials. A nonsimple relationship is suggested by irregularity of the curves. Nonetheless, the "rated" dielectric constant of the material is approached at high values of W/H.
Communications Transistor Corp.

Characteristic impedance of transmission lines   181

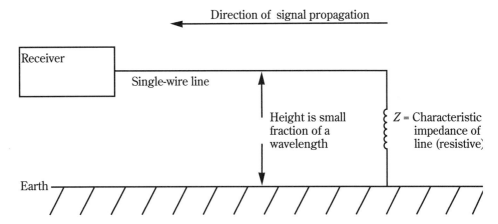

5-24 Unidirectional signal response via a simple beverage antenna. Actually a leaky transmission-line, the Beverage antenna is a large-scale version of one form of a directional coupler.

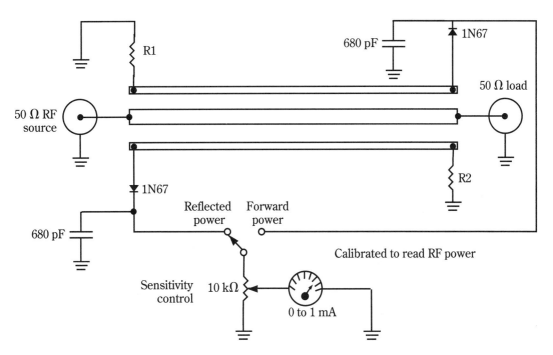

5-25 A typical reflectometer RF wattmeter. Because such an instrument reads forward and reflected power, an accompanying chart converts it into a VSWR indicator.

quency dependent. Because of this inherent behavior, a sensitivity or set adjustment is provided on the front panel. By first setting this adjustment for either a full-scale or reference-point readout when sampling forward power, a correctly calibrated indication of reflected power can then be obtained.

The circuit of Fig. 5-25 is typical of units that are intended for service in 50-Ω systems throughout the 1.8- to 30-MHz range. Precision commercial resistors R1 and R2 will vary according to the physical construction of the line and sensing elements. Values between 33 and 120 Ω are commonly found. The accuracy of the device depends more on the match of these resistors than on their absolute values. Most expensive versions of this instrumentation technique are made to be used in either 52- or 73-Ω systems. Less costly indicators are compromised to provide reasonably accurate readings at either impedance level. This makes sense because these instruments are often used in a relative way, rather than to provide readings of absolute accuracy.

Many variations of these instruments have appeared on the market. Two-meter versions for simultaneously indicating forward and reflected power are quite commonly encountered. Some two-meter types indicate VSWR on one meter and power on the other. More recently, single cross-needle meter movements have been used to enable the operator's glance to encompass forward power, reflected power, and VSWR, while confining attention to system adjustments (Fig. 5-26). Bar-graph readouts are still another example of diversity in commercial instruments. Increasingly popular are RF wattmeters and VSWR indicators with peak-reading circuitry that enables meaningful measurements to be made with SSB transmitters. The basic arrangement of Fig. 5-25, despite some inconveniences will, however, provide useful data to optimize the performance of most RF systems.

The Bruene circuit reflectometer instruments Another family of reflectometer instruments is based on a different type of directional coupler. The circuit shown in Fig. 5-27 uses a current transformer to accomplish selective response to forward and reflected traveling waves on a transmission line. This sampling technique has the advantage of not being frequency sensitive—a VSWR indicator or directional wattmeter using this scheme can readily accommodate a 15:1 frequency range with

5-26 An antenna tuner with a single cross-needle meter. One glance is sufficient to read forward power, reflected power, and VSWR. Moreover, there is no set or sensitivity adjustment to make.

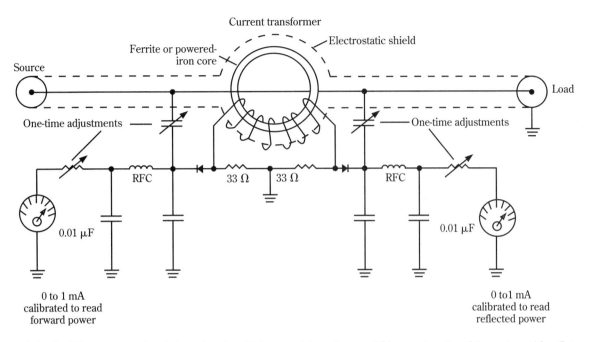

5-27 An RF wattmeter that is based on toroidal current transformer. This constructional format provides flat response over a wide frequency-range.

no need for re-calibration. This enables instruments that use this principle to be factory-calibrated to read power, and to dispense with the need for resetting the meter sensitivity when changing frequency. A drawback of these instruments is that, despite their flat response over a large frequency range, the technique encounters high-frequency limitations; the transformer secondary will approach self-resonance somewhere in the VHF or UHF region if someone attempts to extend operation beyond the HF amateur bands.

The implementation of this directional coupler is very much as indicated in the schematic diagram. The center conductor of a coaxial line is simply passed through the center of the toroidal current transformer and act as a single-turn primary. The secondary might consist of 40 or so turns of fine wire; a compromise must be made between sensitivity and the highest frequency for accurate operation. In any event, as its name indicates, this transformer samples current in the line. *Voltage sampling* is accomplished by the capacitor divider networks. Thus, the diodes are impressed with two-voltages: one that is proportional to the current in the transmission line, and another that is proportional to the voltage at the line.

Directional sensitivity is predicated on a subtle behavior characteristic of transmission lines. Current and voltage in the forward traveling wave are in phase, whereas current and voltage in the reflected traveling wave are 180° out of phase. Thus, the two voltages impressed on the forward-sensing diode will combine additively, whereas the two voltages impressed on the reflection-sensing diode will combine subtractively. Notice that the two sampling diodes experience different phase conditions because the "push-pull" arrangement of the transformer secondary circuit causes the voltage at one end of the secondary winding to be 180° displaced in phase, with respect to the other end. This is more obvious in some models, in which the transformer secondary has a grounded center tap. The metering formats described for the monomatch reflectometers also apply to the Bruene family of similar-purpose instruments.

5-28 The Bird Model 43 directional RF wattmeter. VSWR can be determined by a chart, or by calculations that involve forward and reflected power readings.

Characteristic impedance of transmission lines 185

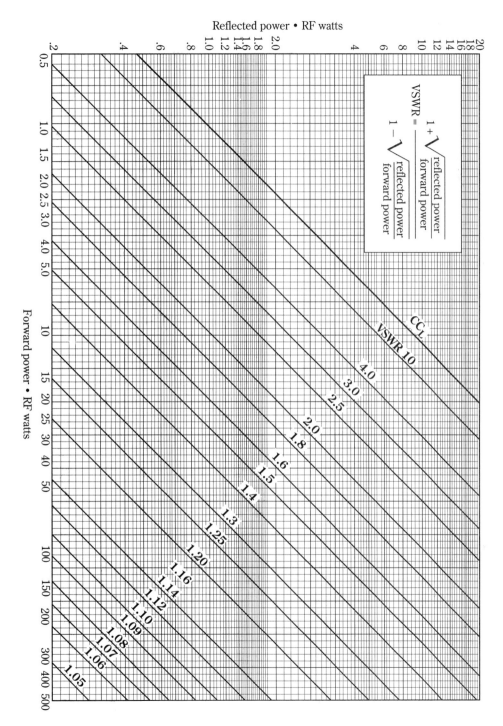

5-29 A chart for determining VSWR from forward and reflected power readings. The indicated formula is also convenient to use, particularly when a calculator is available.

The Bird Thruline RF wattmeters No discussion of VSWR or of directional RF wattmeters could do justice to the subject without mention of the Bird Model 43 Thruline RF wattmeter (Fig. 5-28). This instrument has served as the reference measuring device for many years. During any measurement process, forward and reflected power are indicated on the same meter by simply rotating the plug-in slug 180°, whereon the arrow depicts whether forward or reflected power is being indicated. Once these two parameters are recorded, VSWR can be determined by means of a chart or by calculation (Fig. 5-29). Also, the actual power being delivered to the feedline that is connected to the load is simply the difference between the forward and reflected power readings.

A number of different slugs are available to cover different frequency ranges and power levels. The instrument is internally compensated so that substantially flat response is obtained for a wide frequency range. Later meters, such as the Models 4314 and 4410, incorporate additional features, such as peak-reading capability, switchable power ranges (so that fewer slugs are required), and an RF-monitoring provision to accommodate a frequency counter. These instruments are generally advertised for application in 50-Ω systems, and to cover frequencies from below the broadcast band to the edge of the microwave spectrum. Depending on frequency, power levels from a fraction of a watt to 10 kW can be accommodated. As suggested by the logo, "Thru-line," the insertion loss is negligible.

These instruments conveniently allow absolute power readings to be measured, whereas most VSWR indicators are best used for relative measurements. It is interesting to contemplate also that for most practical purposes, the ratio of forward to reflected power, quickly obtained with the Thruline wattmeters, yields the sought impedance-match information. It is not altogether necessary to directly measure VSWR.

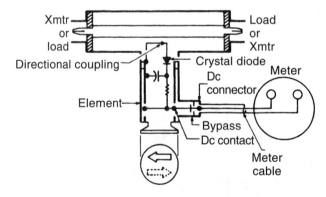

5-30 Basic constructional features of the Thruline slugs. These plug-in units enable the instrument to be used over different frequency bands and power ranges. The rotational feature allows quick measurement of forward and reflected power. Bird Electronic Corp.

The basic constructional features of the slugs are shown in Fig. 5-30. Because of the electrical symmetry of the sampling technique, the source and load connections of the meter can be reversed. In any case, the arrow on the slug will indicate the direction of power flow.

Table 5-2. The standard elements for the Bird Model 43 Thruline directional wattmeter.

Power range (W)	Frequency bands (MHz)					
	2-30	25-60	50-125	100-250	200-500	400-1000
5	—	5A	5B	5C	5D	5E
10	—	10A	10B	10C	10D	10E
25	—	25A	25B	25C	25D	25E
50	50H	50A	50B	50C	50D	50E
100	100H	100A	100B	100C	100D	100E
250	250H	250A	250B	250C	250D	250E
500	500H	500A	500B	500C	500D	500E
1000	1000H	1000A	1000B	1000C	1000D	1000E
2500	2500H					
5000	5000H					

Table 5-2 lists the standard slugs or elements for this instrument. These cover frequency ranges and power ratings that tend to be mostly in demand. However, lower (1 W) and higher (10 kW) power ratings are also available. Also, lower (0.45 to 2.5 MHz) frequency ranges and higher (2200 to 2300 MHz) frequency ranges are available. The general trend in these specifications is that lower power ratings and higher frequency ranges go together. This, agrees with the situation most likely to be found in practice—higher-frequency systems generally operate at lower power levels than lower-frequency RF systems.

Use of the SWR analyzer for quick and convenient RF measurements

A particularly versatile instrument for practical RF work is shown in Fig. 5-31. This particular *SWR analyzer*, is the MFJ Model 247. This self-contained, battery-operated product is of hand-held dimensions and requires minimal time, effort, or skill to obtain precise measurements. It is actually several functional blocks in one: a variable-frequency RF source (covering 160 through 10 m), a 10-digit frequency counter, directional couplers for sensing forward and reflected RF voltage, and a calculator (for converting this data to the VSWR displayed on the SWR analog meter). Notice that this instrument is not intended to be left in the line, but is used to determine the characteristics of antenna's or other loads, feedlines, antenna systems, antenna tuners, transmatches, impedance-transforming networks, baluns, and amplifiers, when these RF components are in their passive states (i.e., the transmitter or RF generator is turned off).

In all cases, automatic readings are produced and no calibration procedure is required beyond the simple tune and search operation. A plug-in adaptor can be acquired from the factory for operation from the 120-Vac line.

- Direct measurements can be conveniently made right at the antenna itself and quickly find the resonant frequency, the 2:1 SWR bandwidth, the proper tuning of traps in multi-band antennas, the behavior of baluns, and other per-

5-31 The MFJ-247 SWR analyzer, a versatile RF measuring instrument. Handheld and battery operated, this instrument is exceptionally convenient to use.

tinent antenna information (such as the effect of orientation or objects). Physical modification of antenna parameters can be instantly monitored for desired and undesired effects.

- The self-resonant frequency of the antenna can actually be determined from observing minimum SWR at the sending end of the antenna feedline with no transmatch in the system.
- A transmatch or antenna tuner can be adjusted without putting the transmitter on the air. Thereafter, the conventional in-line VSWR indicator can be used to reveal any change in the RF system, such as a change in antenna characteristics, feedline defects, etc. The use of the SWR Analyzer eliminates the possibility of damaging the transmitter or the transmatch during the adjustment process, and, of course, accomplishes the desired match without interfering with other station operators. Be sure to remove the SWR Analyzer before turning the transmitter on.

A relevant precaution, as in the use of ordinary in-line VSWR indicators: It is highly desirable for no antenna currents to be on the outside of the coax feedline. The presence of RF on the outer surface of the feedline braid actually makes the feedline part of the radiating antenna and can upset any measurements based on VSWR. This accounts for the years-long controversy over whether the length of the feedline has any pronounced effect on VSWR at the transmitter end of the line. The simple answer is "no" if the braid of coaxial cable is at ground potential. The answer, unfortunately is "yes" if RF is on the coaxial-braid outer surface. RF antenna-currents are ordinarily kept off the outer surface of the braid by using a balun, by having an RF choke composed of ferrite beads slipped over a portion of the line near the antenna, and by keeping the line physically perpendicular to the antenna.

Once the above precaution has been observed, do not worry that off-resonance operation of the antenna or impedance mismatch at the feedline/antenna junction can produce RF antenna currents on the outside of the feedline—this alleged cause/effect relationship is a commonly believed fallacy.

In many practical antenna situations, some RF on the outside of the coax feedline will still allow VSWR determinations to be made with the provision that the readings are then more relative than absolutes, and that they pertain to the antenna system in which the feedline partially behaves as extension of the antenna proper.

It is, indeed, not too far-fetched to expect that on-air tuning of the transmitter and/or the antenna tuner will eventually be made illegal and that all such adjustments must be made with dummy loads and with an instrument, such as the SWR Analyzer.

- The effects of whip-antenna or loading-coil adjustments in mobile installations can be quickly ascertained throughout the band without energizing the transmitter.
- The 10-digit frequency-counter (responsive to 150 MHz) is instantly available for uses other than with the internal SWR and frequency-generator circuits.
- Other models with various modifications are available. For example, the model MFJ 207 SWR Analyzer dispenses with the frequency counter. Instead, frequency is read from a calibrated dial. Often, the extreme accuracy of a digital frequency counter is not needed in practical tune-up situations and the omission of the digital frequency counter saves about half the cost. Then, too, there is a VHF SWR Analyzer, Model MFJ-208, which provides measurement capability over the 142- to 156-MHz frequency range; dial calibration is relied on for reading the frequency. Incidentally, it is never too late to associate a high-capability (to 600 MHz) digital frequency counter with an SWR Analyzer using dial-calibrated frequency readout. MFJ Enterprises makes a 10-digit frequency counter, the Model MFJ-346, specifically for such a purpose.

6
Low-power applications

IN PRACTICE, TRANSISTORIZED CIRCUITS WITHIN THE 1- TO 25-W POWER RANGE, are closer to tube circuit practice than is the case at higher power levels. Also, it is often the case at VHF, UHF, and microwave frequencies that a few watts are sufficient for adequate performance. Yet another important aspect of low-power circuits is that the oscillators, buffers, drivers, and frequency multipliers that precede higher-power final amplifiers operate at much lower power levels. Thus, low-power RF stages are of interest; both as RF output circuits and as building blocks for more elaborate systems.

The philosophy underlying the circuit descriptions is that the reader will most likely establish relevancy with a somewhat similar, but not identical, application. Therefore, this section covers basic principles, rather than detailed, "how-to-build" instructions.

Microwave doppler radar system

The CW transmit-receive system shown in Fig. 6-1 can be applied for speed measurement or intruder-detection functions. Because of the *doppler effect*, a moving object alters the frequency of the reflected energy. The received and transmitted frequencies then heterodyne in the nonlinear characteristic of the oscillator diode, thereby producing an intermediate frequency that is proportional to the speed of the object. A true calibration of the object's speed is usually based upon the premise that the object is moving directly along the line of sight—either toward or away from the radar. Police radar detection of speeding autos is a classic example of this principle.

The bulk-effect diode uses a negative-resistance region, which develops in gallium arsenide semiconductor material when it is suitably biased by a direct current. The oscillation frequency is primarily governed by the associated cavity. However, it is also important that the power supply be ripple-free and well regulated in the interest of optimum frequency stability.

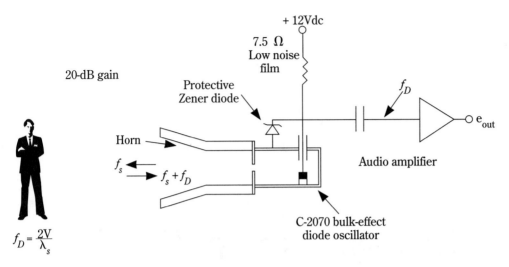

6-1 Speed measurement or intrusion-detection radar using bulk-effect diode. The diode and cavity are available as an integral unit. _{General Electric Microwave Devices Products Section}

The diode used in this doppler radar is integrally associated with its own oscillator cavity and operates at 10,525 MHz (see Fig. 6-2). It is only necessary to attach a horn, and connect the bias source and a suitable audio amplifier. If the audio passband extends from 100 Hz to 3 kHz, the output tone will represent speeds that range from approximately 3.33 to 100 miles per hour. Either analog or digital circuit techniques can be used to produce an appropriate meter readout or to actuate an alarm.

Notice that in this application, the horn antenna is effective for both outgoing and incoming microwave energy. Practical horns can provide 16.5 dB gain. You can, therefore, deal with an effective radiated power of 33 dB greater than would be gained from coupling the oscillator to a simple dipole radiator in an ordinary transmitting application. This is a typical example of how low power levels can be put to practical use in microwave applications. Working distances up to several hundred feet are feasible. The reflecting characteristics of the target and the noise characteristics of the oscillator diode are the main distance-limiting factors. The gain of the audio amplifier is important, too, but high gain is only useful if the doppler IF signal is not masked by oscillator noise.

1-MHz JFET crystal oscillator

The JFET circuit shown in Fig. 6-3 is a general-purpose crystal oscillator. Its salient feature is simplicity. It is useful where much better stability than is readily forthcoming from self-excited oscillators is desired. On the other hand, it is not recommended for applications where the best potentialities of crystal control are sought. Although a p-channel JFET is shown, n-channel devices will work equally well, providing that the polarity of the supply voltage is reversed. This circuit is the JFET ver-

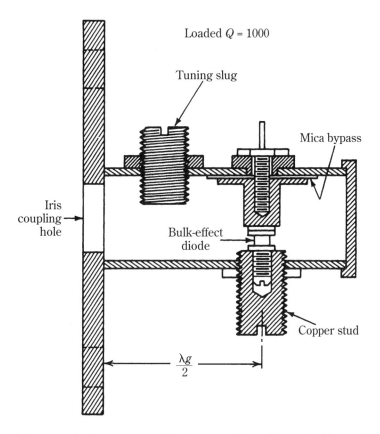

6-2 Detailed view of diode-cavity assembly. Operation is readily attained by attaching a horn and connecting bias source and audio amplifier. <small>General Electric Microwave Devices Products Section</small>

sion of the Miller oscillator of former popularity with tubes. By making appropriate changes in the LC resonant tank, other frequencies can be generated with appropriate fundamental-frequency crystals. Unreliable starting performance can usually be remedied by connecting an external feedback capacitor of a few picofarads from drain to gate.

Because crystal oscillators are generally followed by one or more amplifier stages, the low power output of small JFETs do not need to be a disadvantage. In any event, a buffer amplifier is very desirable in order to reduce loading and to minimize frequency pulling. Unfortunately, it has no internal buffering, such as existed in tetrode and pentode tube versions of the circuit. Nonetheless, it is a commonly encountered workhorse and is a satisfactory performer in a variety of applications. In particular, the high-impedance gate circuit enables it to support the parallel-resonant mode of oscillation in medium-frequency crystals. For best results, the subsequent buffer stage should also be a JFET. Although the output is available directly at the drain terminal, as shown, it will generally be better to use an approximate mid-tap on the inductor in conjunction with a small series capacitor.

194 Low-power applications

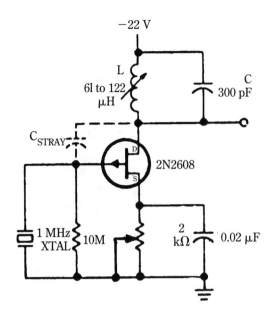

6-3 A JFET version of the Miller crystal-oscillator circuit. This general circuit has served as a workhorse for numerous applications where the demands for stability and frequency precision are moderate. _{Motorola Semiconductor Products, Inc.)}

JFET frequency doubler

The JFET frequency-doubling circuit of Fig. 6-4 is designed for low-power operation, but the power level can be scaled up somewhat via appropriate device selection and a higher supply voltage. The operating efficiency and wave purity of this doubler is much higher than is readily attainable from conventional frequency multipliers. The input circuit is push-pull, whereas the output circuit is parallel. Such a configuration favors the support of even-order harmonics in the output and, at the same time, tends to cancel odd harmonics. Because the fundamental frequency is, itself, an odd harmonic, relatively little wave-purity contamination can occur from feedthrough of the fundamental in this doubler.

The most interesting aspect of this doubling scheme has to do with the mechanism of second-harmonic production. This is because of the unique transfer characteristic of JFETs. Specifically, it approaches a true square-law response. The use of the grounded-gate connection for the JFETs results in more practical resonant-circuit components, but the experimenter can adapt the basic idea for grounded-source operation. As shown, input frequencies in the vicinity of 60 MHz are doubled. The output trap, L2/C8 attenuates residual third-harmonic energy in the output. By means of this technique, together with matched JFETs, third-harmonic rejection can be as high as 70 dB below $2f$. However, many practical applications will allow the trap to be dispensed with because 50-dB rejection of $3f$ can still be obtained.

Unlike most active-device frequency multipliers, the doubled frequency in this system represents the main distortion product in the JFETs. That is, $2f$ in the output is not produced by shock-excitation of the resonant output circuit. To optimize such second-

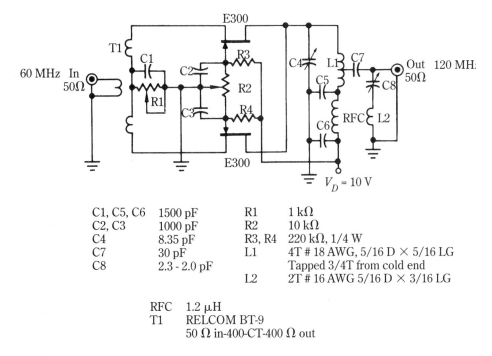

6-4 High-efficiency JFET frequency doubler. The JFET transfer characteristic and the circuit configuration contribute to enhancement of second-harmonic production. _{Siliconix Corp.}

harmonic distortion, variable resistance R, is included so that the most effective bias can be empirically determined. Because of the range provided by R1, the pinchoff voltage of the selected JFETs is not of great consequence. In any event, R1 is adjusted to optimize the frequency-doubling characteristic of the circuit. The balancing potentiometer, R2, is adjusted for minimum presence of the fundamental frequency in the output. To a considerable extent, R2 will compensate inequalities in the two JFETs.

Class-C amplifier: 1.5 W at 50 MHz

An interesting aspect of the amplifier shown in Fig. 6-5 is its configurational resemblance to class-C tube amplifiers. For example, the variable resistance, R, serves the same function as the "grid leak" or grid return resistance in a self-biased tube circuit. C6 is the counterpart of the "grid capacitor" in such circuits. That is, under drive conditions, a reverse-bias charge is trapped in C6, and this charge biases the transistor well into the class-C operating region. In practice, the drive power and R can be mutually adjusted to achieve maximum RF output from the amplifier. Although this will correspond to a value of R that is greater than zero, such biasing is not generally encountered in transistor RF amplifiers—at least not in those with greater power capability than this amplifier. The reason is that most RF power transistors become more vulnerable to catastrophic breakdown as reverse bias is increased. However, the technique is feasible for low-power output amplifiers and drivers. Probably 5 W or so is the peak level; it might not be wise to operate deeper into class-C than naturally ensues

196 Low-power applications

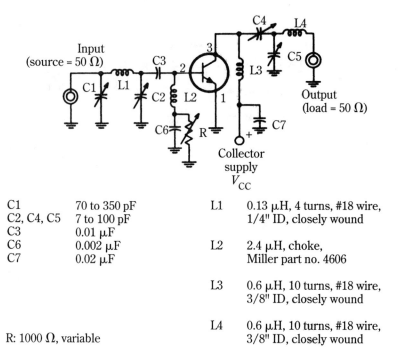

C1	70 to 350 pF	L1	0.13 µH, 4 turns, #18 wire, 1/4" ID, closely wound
C2, C4, C5	7 to 100 pF		
C3	0.01 µF		
C6	0.002 µF	L2	2.4 µH, choke, Miller part no. 4606
C7	0.02 µF		
		L3	0.6 µH, 10 turns, #18 wire, 3/8" ID, closely wound
R: 1000 Ω, variable		L4	0.6 µH, 10 turns, #18 wire, 3/8" ID, closely wound

6-5 Class C amplifier: 1½ watt, 50 MHz. This simple circuit provides a convenient way to explore solid-state VHF techniques. _{RCA Solid-State Division}

from directly grounding the base RF choke. Aside from endangering the transistor, this biasing technique often causes the emitter-base diode of the transistor to go into Zener or avalanche breakdown, which thereby defeats the self-bias function and sometimes causes loading problems in the driver stage. In this amplifier, the base RF choke, L2, has a low Q. This discourages low-frequency "parasitic" oscillation.

The RCA 2N3118 transistor can be safely used in this circuit with a 40-V dc supply. Under such conditions, 60 mW of drive power is sufficient for 1½-W output power. With 120 mW of drive, 2-W output can be approached. Sufficient heatsinking should be applied to the TO-5 case to keep it near 25°C during operation.

The internal feedback capacitance of this transistor is sufficiently low that neutralization is not needed. However, this does not obviate the usual precautions that pertain to short leads, input-output isolation, and effective bypassing. With respect to the latter, it might be profitable to parallel C7 with both a smaller and a larger ceramic capacitor.

This amplifier can be readily adapted for either CW or FM service in the amateur 6-m band. It is not, however, suitable for SSB, and it lacks the peak voltage and power capability to be amplitude modulated.

JFET broadband linear amplifier

The inordinately simple amplifier circuit shown in Fig. 6-6 is "linear" in the sense that it is eminently suitable for processing AM signals. It is intended for insertion of a 75-Ω system, such as is commonly used in CATV and MATV systems. However, it is

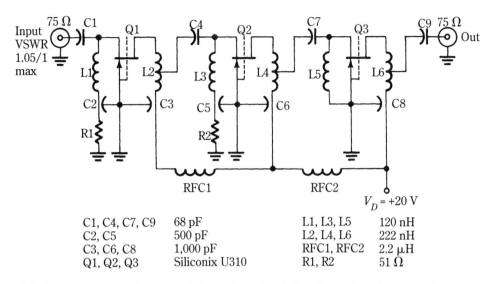

6-6 JFET linear amplifier for the 200–250-MHz band. Simplicity of construction is the salient feature of this broadband amplifier. _{Siliconix}

likely that little or no modification would be needed for satisfactory 50-Ω operation. Although the input, output, and interstage networks border on the primitive, the response is within ±1 dB throughout the 200- to 250-MHz band. The low-input VSWR is achieved because JFETs can display virtually no input reactance when connected in the grounded-gate configuration.

Some experimentation is needed to find the optimum taps on inductors L2, L4, and L6. Center tapping is a good start for such empirical investigation. Reasonably good performance will be immediately forthcoming, and the taps are not critical. Conspicuous by their absence in this circuit are resonating capacitors. Also missing, but not missed, are bias networks or bias sources. Some effort should be directed toward placing a shield partition between source and drain leads. This prevents positive feedback in the manner of ultraaudio or Colpitts oscillators. The intermodulation and power-level performance of this JFET amplifier is graphically depicted in Fig. 6-7.

Power MOSFET driver for a broadband linear amplifier system

The circuit shown in Fig. 6-8 is intended as a basic "building block" for a more elaborate linear amplifier. In particular, it would be suitable for driving a power MOSFET output stage. The adage "a chain is no stronger than its weakest link" appropriately applies to a lineup of linear-amplifier stages. If a driver stage is nonlinear and thereby develops high intermodulation distortion, all design efforts expended on the output stage for the sake of system linearity are in vain. This driver attains its low intermodulation distortion by virtue of the inherent transfer linearity of power MOSFETs when they are properly biased. As can be seen, provision

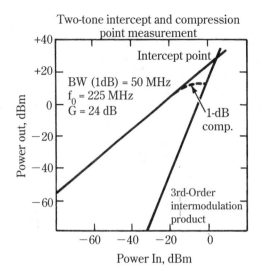

6-7 Performance of the JFET linear amplifier of Fig. 6-6. Siliconix

is made for the introduction of the bias voltage, V_{GS}. Although a variable source of positive voltage is needed (approximately 4 V), virtually no bias current is consumed by the gate.

The transmission-type transformer, T1, resembles types that are commonly used as baluns for converting from balanced to unbalanced circuits, or vice versa. However, in this instance, it is the impedance transforming property of T1 that is exploited. Specifically, the 50-Ω input impedance is transformed to ¼ of this value, 12.5 Ω, in order to achieve a reasonable impedance match at the gate throughout the 40- to 265-MHz range. The near-infinite input impedance often cited as a feature of the power MOSFET prevails only at dc and for very low frequencies. Another factor in producing the low gate impedance is the LR feedback network between the output and the gate.

The inordinately broad response of this amplifier is achieved from the combined action of transformer T1 and the LR feedback network. The 0.15-µH feedback inductor can be made by winding about seven turns of #30 enamel wire on a ½-W, 1-MΩ resistor. Molded inductors will probably have too much distributed capacity to be useful for this function. The effect of such an inductor will reduce the upper-frequency bandwidth. The gain and intermodulation performance of this circuit is depicted in Fig. 6-24, along with that of a two-transistor version.

Low-power 2-M linear power MOSFET amplifiers

Simplicity, together with respectable performance, characterizes the 2-M linear amplifier that is shown in Fig. 6-9. Simple lumped-circuit resonant circuits are used, and there is no need for neutralization or feedback networks. R5 enables adjustment of

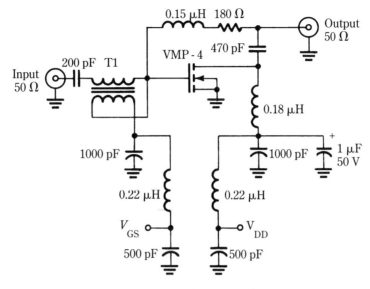

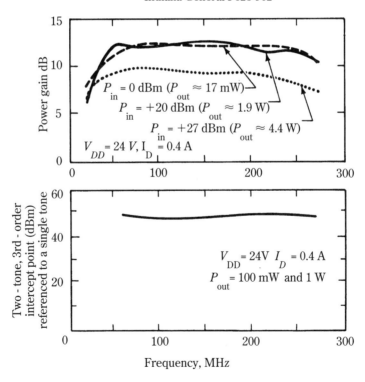

6-8 Simple 2-W broadband driver for a linear amplifier system. Although no tuned circuits are used, the intermodulation and harmonic distortion are acceptably low. _{Siliconix Corp.}

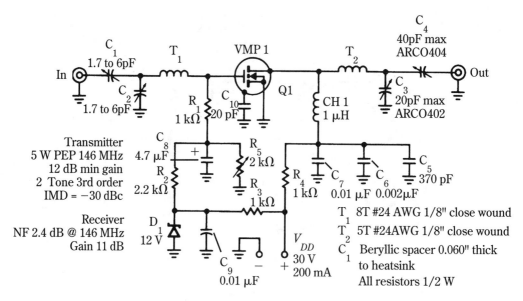

6-9 A 2-m, 5-W PEP linear amplifier that uses a power MOSFET. Although intended as a transmitter output amplifier, the very same circuit surprisingly performs as well as an RF preamplifier for a receiver. Siliconix Corp.

the bias for optimum linearity. Unlike a bipolar transistor, the power MOSFET does not consume dc bias current. To promote stable and reproducible operation, the bias supply uses a 12-V Zener diode. The input and output impedances are 50 Ω. The power MOSFET can withstand any VSVR, so no precautions are necessary during adjustment or operation. Considerable inherent protection from overdrive or overloading exists because the output current decreases with rising temperature. Because the case of the device is internally connected to the drain, the heatsink must be insulated from the ground plane of the circuit board. The VN66AJ power MOSFET can be used with substantially the same results.

Another version of this basic amplifier is shown in Fig. 6-10. Here, two power MOSFETs contribute nearly twice the output power available from the single device amplifier. The two power MOSFETs are essentially in parallel, with respect to the RF. However, they are not connected in a brute-force parallel arrangement, such as would result from connecting their respective terminals together. The scheme used enables individual adjustment of optimum bias linearization of the two power MOSFETs. Additionally, better harmonic rejection is obtained than would be the case if the simpler paralleling technique were used. The inductors are the same as the corresponding ones in the single device circuit of Fig. 6-9. The 1-Ω current-sense resistors in these amplifiers enable drain current to be monitored with commonly available milliammeters or voltmeters.

135-MHz AM amplifier

The three-stage amplifier in Fig. 6-11 is suitable for use in the aircraft communications band, where AM has prevailed over FM and SSB modulation modes. The am-

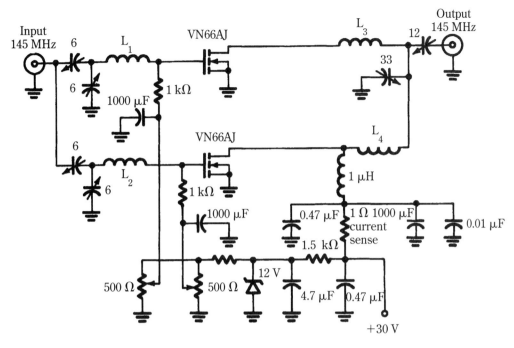

6-10 A 2-m, 10-W PEP linear amplifier that uses a pair of power MOSFETs. The indirect paralleling technique essentially doubles the PEP power that is available from the single-device circuit of Fig. 6-9. <small>Siliconix Corp.</small>

plifier provides 6 W of output, and its upward modulation factor exceeds 90%. Some explanation is in order here, for the circuit is deliberately designed to enable a higher percentage of modulation to be achieved in the upward (increasing output) than the downward direction. Such an asymmetrical modulation envelope naturally produces distortion and certainly would not be suitable for the high-fidelity transmission of music. However, for voice transmission, its effect on intelligibility is quite benign, and is overshadowed by the effective "talk-power" that accompanies high-percentage modulation.

The salient advantage of such asymmetrical modulation is that practically full-modulation capability can be realized without danger of overmodulation, with its resultant splatter in the RF spectrum. Whereas excessive modulation in the downward direction is serious, upward modulation can even exceed 100% without such deleterious effects. Indeed, radio amateurs at one time used "exalted" modulation techniques to purposely "overmodulate" in the upward direction. This was a popular way to obtain some of the features of higher power, but at low cost. Signal splatter did not occur as long as the carrier level was not downward modulated excessively. However, too much upward modulation ultimately produces unacceptable distortion in the detectors of some receivers.

Notice in Fig. 6-11 that the second and third stages of the amplifier are conventionally modulated. Such tandem modulation is commonly encountered in AM solid-state amplifiers. The reason is that it is difficult to achieve high modulation factors

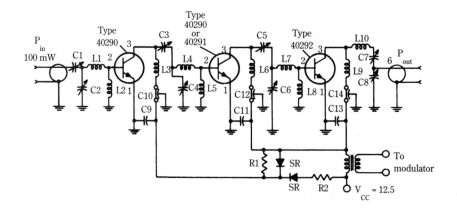

C1, C3, C5, C7,	3 to 35 pF	L1, L9	3 turns #16 wire, 1/4" ID, 1/4" long
C2, C4, C6, C8	8 to 60 pF	L2, L5	Ferrite choke, $Z = 450\ \Omega$
C9, C11, C13,	0.03 µF	L3	RF choke, 1.5 µH
C10, C12, C14,	1000 pF	L4, L7	4 turns #16 wire, 1/4" ID, 3/8" long
		L6	RF choke, 1.0 µH
R1	220 Ω	L8	Wire-wound resistor, $R = 2.4\ \Omega$
R2	180 Ω		
SR	1N2858	L10	5 turns #16 wire, 3/8" ID, 1/2 long

6-11 Amplitude-modulated amplifier for 135 MHz. Suitable only for aircraft communications, "talk-power" is enhanced by an asymmetrical modulation envelope wherein upward modulation exceeds downward modulation. Minimal effect on speech intelligibility results, and signal splatter is avoided. <small>RCA Solid-State Division</small>

by applying the audio modulation to the final RF power stage only. If attempted, it is generally found that the modulated transistor saturates well before 100% upward modulation can be attained. By simultaneously modulating the driver stage, the saturation region is moved up to accommodate heavier modulation.

Additionally, the first RF amplifier stage is modulated from audio that is derived from the diode-resistor network connected across the secondary of the modulation transformer. This network permits the first RF amplifier stage to be modulated in the upward direction, but it drastically limits its downward modulation. The overall result is that the RF output of the amplifying system can be safely modulated at high modulation factors without danger of overmodulation in the usual sense. Carrier cutoff and signal splatter cannot occur.

Editor's note: The 135-MHz AM amplifier is included in this book for informational purposes only. The unlicensed use of such equipment is illegal and could cause great harm.

Doubling the output power (almost)

The circuit shown in Fig. 6-12 is substantially the same as that of Fig. 6-11, except that the final stage consists of two parallel-connected power transistors. The paral-

135-MHz AM amplifier

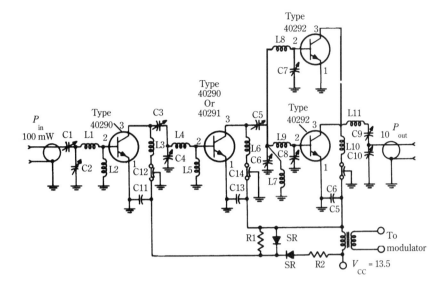

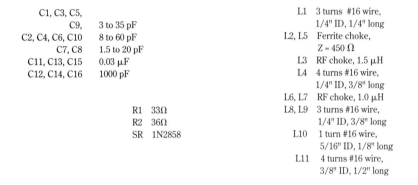

C1, C3, C5, C9,	3 to 35 pF
C2, C4, C6, C10	8 to 60 pF
C7, C8	1.5 to 20 pF
C11, C13, C15	0.03 µF
C12, C14, C16	1000 pF

R1	33Ω
R2	36Ω
SR	1N2858

L1	3 turns #16 wire, 1/4" ID, 1/4" long
L2, L5	Ferrite choke, Z = 450 Ω
L3	RF choke, 1.5 µH
L4	4 turns #16 wire, 1/4" ID, 3/8" long
L6, L7	RF choke, 1.0 µH
L8, L9	3 turns #16 wire, 1/4" ID, 3/8" long
L10	1 turn #16 wire, 5/16" ID, 1/8" long
L11	4 turns #16 wire, 3/8" ID, 1/2" long

6-12 A 10-W, 135-MHz AM amplifier with a paralleled output stage. Paralleling of RF power transistors generally does not involve the brute-force connection of the respective terminals. In this case, individual base-drive networks are used. <small>RCA Solid-State Division</small>

leling arrangement is a bit more involved than merely connecting the respective terminals of the transistors together. Even so, it is difficult to obtain twice the output of a single transistor stage. This is primarily because of differences in the transistors. The situation is somewhat like paralleling two batteries that do not have quite the same voltage, voltage regulation, internal resistance, and temperature behavior. Another complication arises from the doubled power demand that is imposed on the driver stage. Perhaps with RCA 40292 output transistors, you might squeeze 11 W from this arrangement.

Paralleling in higher-power output stages often involves individual networks in the collector, as well as in the base circuits. This tends to be closer to actually doubling the power available from a single transistor. Because of the added complication, such an output stage is less frequently seen than the push-pull connection, which provides the possibility of greatly attenuating even-order harmonics.

Complementary-symmetry driver amplifier for SSB

The broadband (2 to 30 MHz) amplifier in Fig. 6-13 is the equivalent of two cascaded push-pull stages, even though single-ended resonant tank circuits are used. Such an arrangement was not feasible in tube designs because no pnp tube was ever available. The output power is up to 25 W PEP, and the power gain is in the vicinity of 35 dB. As depicted, the amplifier is intended for use as a driver for a larger amplifier. However, with the use of appropriate harmonic filters in the output, this amplifier could also serve as a final power stage to feed an antenna.

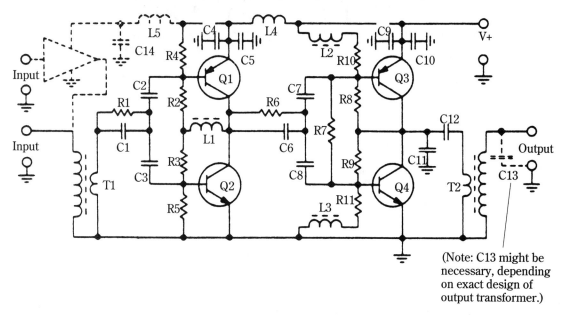

(Note: C13 might be necessary, depending on exact design of output transformer.)

C1, C6	1000 pF	L1	0.68 μH molded choke
C2, C3, C7, C8	0.1 μF	L2, L3	6.8 μH molded choke
C4, C9, C14	0.15 μF	L4	4 ferrite beads, Ferroxcube 56 590 65/38
C5, C10, C12	0.68 μF		
C11	2000 pF	Q1, Q3	MRF432
C13	100 pF (Typ.)	Q2, Q4	MRF433
R1, R4, R5	22 Ω 1/4 W	T1, T2	See text
R2, R3	330 Ω 1/2 W		
R6	15 Ω 1/4 W		
R8, R9	220 Ω 1 W		
R10, R11	6.8 Ω 1/2 W		
R7	560 to 680 Ω 2 W		

All capacitors are ceramic chips, Union Carbide type 1225 or 1813 or Varadyne size 18 or 14 or equivalent.
All resistors are carbon composite.

6-13 A broadband complementary-symmetry amplifier for SSB. This amplifier is capable of supplying a 25-W PEP throughout the 2- to 30-MHz frequency range. It is intended for service as a driver for a larger amplifier. Motorola Semiconductor Products, Inc.

The dashed-line components provide an option, the use of a hybrid amplifier to increase the power gain to 45 or 50 dB. Motorola's MHW570 to MHW572 types are suitable for this application (gain compensation to cause roll-off below 5 MHz should be used). The purpose of this option is to enable the amplifier to be directly driven from a balanced modulator.

The first stage, Q1 and Q2, is biased for class-A operation, with quiescent collector current set at 700 mA. Q3 and Q4, the second stage, are biased for class-AB operation with a quiescent collector current of about 75 mA. This biasing technique results in good even-order harmonic cancellation, as well as acceptable intermodulation distortion. Negative feedback is used with both stages. The feedback loop of the input stage contains the inductor, L1, and thereby produces gain compensation. This is necessary in most broadband systems to offset the tendency of transistors to develop progressively higher gain at lower frequencies.

Transformers T1 and T2 are similar, but do not have the same winding ratios. Both use Stackpole dual balun ferrite cores (number 57-1845-24B). This core material has a permeability of 2400. The low-impedance windings of both transformers consist of one turn of copper braid. The primary of T1 uses two turns of #22 Teflon-insulated stranded wire. The secondary of T2 is formed of four turns of the same type of wire. In both transformers, the wire is threaded through the braid to obtain close coupling. If more convenient, separate ferrite sleeves with the same magnetic characteristics stipulated for the dual balun structure can be used for the cores. In any event, the general construction of such transformers is illustrated in Fig. 6-14.

6-14 Typical ferrite-core RF transformers that are designed for broadband operation. Close coupling is obtained by threading the wire of one winding through the copper braid of the other. Transformers T1 and T2 of Fig. 6-13 are constructed in this manner.

206 *Low-power applications*

The salient operating parameters of this amplifier are depicted by the curves in Fig. 6-15. The component-layout drawings of Fig. 6-16 represent good high-frequency practice and convenient construction.

2-W 2.5-GHz oscillator that uses the common-collector circuit

Amid the controversies concerning the relative merits of common-emitter and common-base circuits for VHF, UHF, and microwave power, special transistors were developed for operation as common-collector oscillators. Although this configuration is

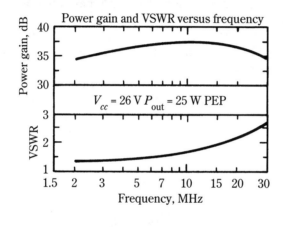

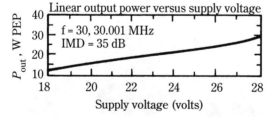

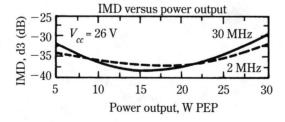

6-15 The operating parameters of the complementary-symmetry driver amplifier. Broadband performance and acceptably low intermodulation distortion make this amplifier usable for amateur SSB applications. Motorola Semiconductor Products, Inc.

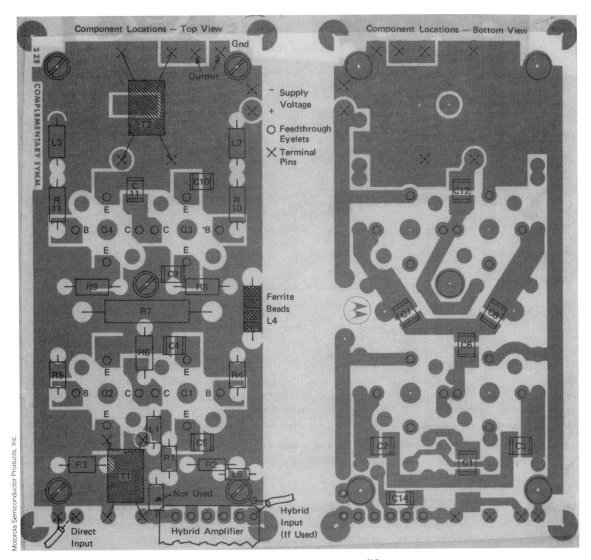

6-16 The component layout for the complementary symmetry amplifier.

not suitable for use as a stable amplifier, it is favorable that the parasitic package capacitances can be optimized to serve as the "voltage divider" of the Colpitts oscillator circuit. Under such conditions, it is feasible to operate the collector at ground potential for both RF and dc, thereby enhancing heat removal and circumventing RF radiation, coupling, and detuning from the heatsink.

Figure 6-17 shows the circuit of a 2-W 2.3-GHz oscillator that uses the TRW 62602 common-collector transistor. The approximate equivalent circuit is shown in Fig. 6-18. Those familiar with tube practice will recognize the configuration as that of the Clapp series-tuned version of the Colpitts oscillator. The voltage-divider capaci-

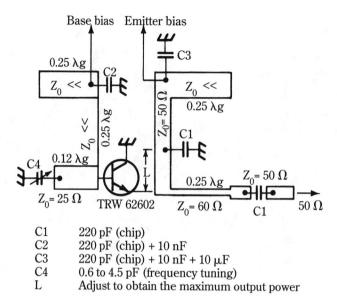

C1	220 pF (chip)
C2	220 pF (chip) + 10 nF
C3	220 pF (chip) + 10 nF + 10 μF
C4	0.6 to 4.5 pF (frequency tuning)
L	Adjust to obtain the maximum output power

6-17 A 2-W, 2.3-GHz oscillator that uses a common-collector circuit. The transistor is designed and packaged to deliver optimum performance in this type of oscillator circuit. TRW

tances, C_{BE} and C_{CE} derive from the transistor and its packaging. Although this concept appears simple enough, it is not always readily applicable with transistors that are not specifically designed for such operation. Although appropriate values for the voltage-dividing capacitors can be attained with external chip capacitors, "ordinary" microwave transistors will not operate with their collectors at RF ground. This is because of the inductance of the collector bonding wires. Although this effect can be somewhat reduced via an external reactance, such implementation is obviously a nuisance.

The circuit of Fig. 6-17 relies heavily on stripline transmission-line elements for simulating the function of tank inductance, RF chokes, the impedance-matching transformer, and the bypass capacitance. In the base circuit, the 25-Ω ⅛-λ composes the series inductor of the frequency-determining tank; variable capacitor C4 adjusts the frequency of the oscillator. The thin ¼-λ line in the base circuit serves as an RF choke, and the wide ¼-λ stud that is associated with it provides effective RF bypassing.

In the emitter output circuit, line-section L cancels the emitter output capacitance. The ¼-λ 60-Ω line is then used to transform the emitter impedance to 50 Ω. The remainder of the structure that contains L serves as RF choke and stub line for bypassing to ground. Because of the copper-dielectric-copper fabrication of the PC board, a ground plane is underneath all of the stripline elements.

Other RF circuitry features are as follows: capacitor C1 provides dc blocking at the output. Two C1s are shown; the other, of identical type and size, is a bypass capacitor. L is governed by the physical location and connection of this capacitor.

The PC board layout is shown in Fig. 6-19. The drawing is not reproduced at full scale. The widths and lengths of the stripline elements are critical. The philosophy of merely approximating geometric shapes and dimensions is permissible for low-

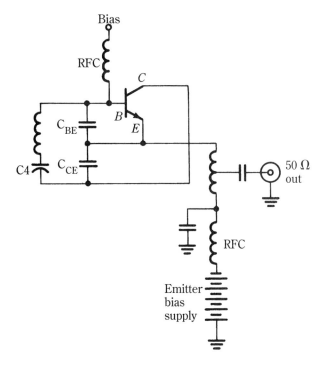

6-18 An approximately equivalent circuit of the microwave oscillator from Fig. 6-17. The circuit is essentially a series-tuned Colpitts (Clapp) oscillator with an impedance-matching output provision. Capacitors C_{BE} and C_{CE} are parasitic packaging capacitances that are put to good use.

frequency electronic circuits, but not for microwave projects. Moreover, it is important to use exactly the prescribed type and thickness of board material. Otherwise, it is likely that a different dielectric constant will be encountered; this will invalidate the dimensions. A microwave "circuit" of this type is as much a geometrical design as it is a connection diagram.

A suitable source of base and emitter bias is shown in Fig. 6-20. This is not the thermal-tracking type that is used with higher-powered RF circuits, but it is adequate for the purpose. It provides adjustment of the operating mode between class B and class A and enables the oscillator to be self-starting. The emitter bias supply corresponds in function to the usual collector supply for common-emitter and common-base circuits. In this application, however, the supply must float, with respect to the ground system of the oscillator. A voltage-regulated supply with a nominal output of 20 V and current capability of 500 mA is recommended for the emitter supply (depicted as a battery in Fig. 6-20).

Tapered-stripline microwave amplifier

The simple and straightforward grounded-base amplifier shown in Fig. 6-21 belies its performance capabilities: 13 W with "tank circuits" that are selected for 1 GHz,

210 *Low-power applications*

PC board layout for F_0 = 2.3 GHz (BW = 500 MHz)

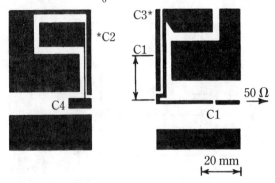

*Foil-wrap asterisked edge to ground plane.
Board material: −0.020" glass Teflon (e_r = 2.55).
Adjust L to obtain the maximum output power.

For F = 2 GHz L = 24 mm
F = 2.3 GHz L = 19 mm
F = 2.5 GHz L = 14 mm

6-19 PC board layout for the 2.3-GHz microwave oscillator. Appropriate deployment of transmission-line elements serves the functions of resonating, impedance transformation, RF choke, and bypassing.   TRW

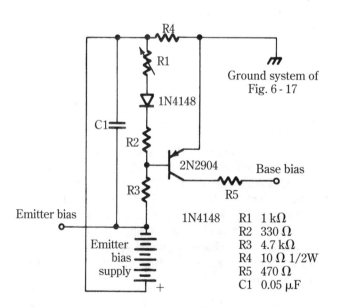

6-20 A base and emitter bias source for the 2-W, 2.3-GHz oscillator. Notice that this dc power supply must "float" with respect to the ground system of Fig. 6-17.

and 6 W when operated with 2-GHz tanks. These input and output tank circuits perform impedance matching via their tapered shapes. The relatively low impedances of both the emitter and the collector circuits are converted to 50 Ω by these tapered stripline elements. No stublines or additional reactances are needed. In principle, the transition from wide to narrow end should be by means of exponential curves. In practice, the simple trapezoidal shape is often sufficient.

The 2N6266 emitter-ballasted transistor operates from a 28-Vdc supply and develops a collector efficiency in the vicinity of 40% at these microwave frequencies. At the same time, it is very rugged, with an infinite VSWR capability. The power

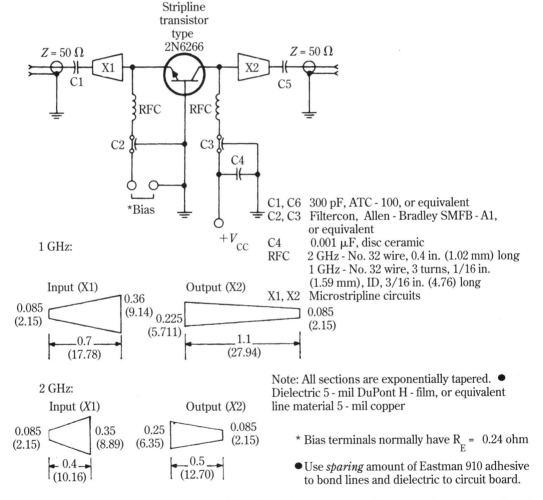

6-21 The tapered-stripline microwave amplifier. Simple impedance-matching networks are suggestive of acoustic techniques. RCA Solid-State Division

levels are quite respectable for the microwave region, and good results can be anticipated for such applications as collision-avoidance systems, S-band telemetry, distance-measuring equipment, microwave-relay links, phased-array radar, and transponders. When incorporated into amplifier chains, driver requirements can be determined from the nominal 11-dB gain that is developed at 1 GHz and 7-dB gain at 2 GHz. The construction of this amplifier is shown in Fig. 6-22.

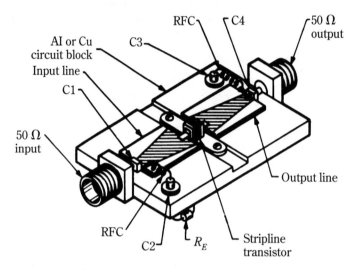

6-22 Construction of the tapered-stripline microwave amplifier. By using appropriate stripline elements, the output power is approximately 13 W at 1 GHz and 6 W at 2 GHz.
RCA Solid-State Division

Warning: The ceramic body of this device contains beryllium oxide. Do not crush, grind, or abrade these portions because the dust resulting from such action could be hazardous if inhaled. Disposal should be by burial.

Quadrature amplifier circuit

The linear amplifier in Fig. 6-23 is essentially a doubled-up version of the circuit in Fig. 6-8. Of prime interest here is the combining technique, which uses of coupled transmission-line elements. The bandwidth is not as great as in the simpler single-device arrangement, but it still qualifies as broadband (100 to 160 MHz). Although gain and output penalties are exacted by the 50-Ω "dummy" loads in the input and output circuits, considerable gain is made in the linearity of the amplifier. In this respect, such an amplifying system is superior to the parallel or push-pull combining techniques. The quadrature combining scheme uses transmission-line elements that are much shorter than a quarter-wavelength. Such short lines impart a 90° phase shift to an introduced signal. By appropriately cross-connecting the ter-

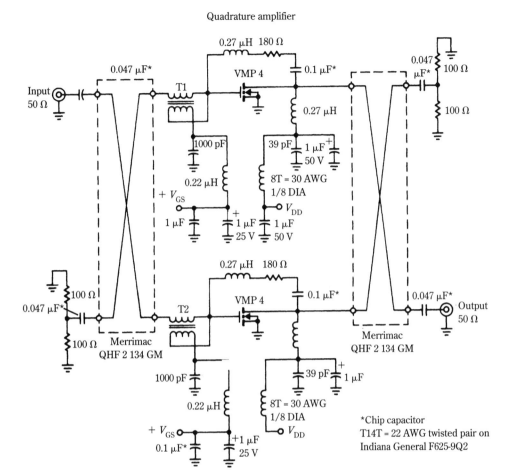

6-23 A quadrature amplifier for 100- to 160-MHz (linear service). Power splitting at the input and power combining at the output are accomplished via coupled transmission-line elements.   Siliconix Corp.

minations of these lines, the same 180° phase displacements that exist in a push-pull amplifier are obtained. Do not confuse this arrangement with a diplexer combining technique, which also uses transmission-line elements. The operation of the diplexer depends on the ¼-λ characteristics of the lines and is therefore a narrow-band scheme.

The provisions for individual gate-biasing help to adjust for optimally low intermodulation distortion. Figures 6-24A and B compare gain and intermodulation performance of this amplifier with that of the single-device circuit of Fig. 6-8. The quadrature amplifier provides better linearity, although it sacrifices about 3 dB of gain.

214 Low-power applications

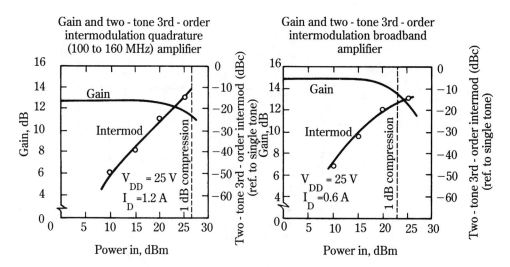

6-24 The gain and intermodulation performance of the amplifiers in Figs. 6-23 and 6-8. A. Two power MOSFETs in the quadrature circuit of Fig. 6-23. B. A single power MOSFET in the broadband circuit of Fig. 6-8. _{Siliconix Corp.}

7
Medium- and high-power applications

THIS CHAPTER COVERS PRACTICAL CIRCUITS IN WHICH THE OUTPUT POWER level exceeds 25 W. Power-frequency combinations that, until recently, were "blue-sky" concepts, are shown as candidates for everyday breadboarding experiments or as well-engineered circuits for ready implementation into equipment. This should be a dynamic field during the next decade, with the major semiconductor firms vying with one another to win the competitive edge in power and frequency capability, efficiency, reliability, and cost.

32-W marine-band amplifier

The 156-MHz amplifier in Fig. 7-1 is intended for FM service. This greatly relaxes some of the stringent demands that other modulation formats impose on amplifiers. The final stage has two 2N5996 transistors operating in parallel. The paralleling configuration is somewhat similar to one or two that were previously featured. However, in this case, the two output transistors each have their individual networks. In the arrangement, the transistors can be made to share equally in power amplification—even though they do not exhibit exactly the same input and output impedances. It would be only natural to ponder why a push-pull circuit is not used because the circuit complexity and parts count would be about the same. The primary reason is that the salient feature of push-pull amplification (cancellation of even-order harmonics) is not always readily forthcoming in practice. To fully exploit the potentialities of push-pull operation, it is generally desirable to use transmission-line tank circuits at VHF frequencies and to pay heed to factors that affect balanced operation. For ex-

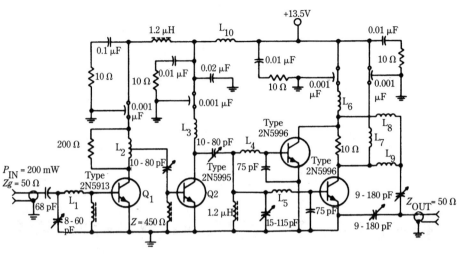

- L_1 2 turns of #20 B.T., 1/4" diameter, 1/8" long
- L_2 5 turns of #20 B.T., 1/4" diameter, 3/8" long, tap at 4 1/2 turns from collector
- L_3 5 turns of #20 enamel wire, 3/16" diameter, 1/4" long
- L_4, L_5 1 turn of #20 B.T., 1/8" diameter, 1/8" long
- L_6, L_7 2 turns of #20 B.T., 3/16" diameter, 1/4" long
- L_8, L_9 2 turns of #18 B.T., 1/4" diameter, 3/16" diameter
- L_{10} 10 turns of #20 enamel wire, 1/4" diameter, close wound

7-1 A 32-W marine-band amplifier. Unique paralleling technique accommodates the parameter tolerances of power transistors.   RCA Solid-State Division

ample, divergent transistor characteristics are more difficult to cope with than in the parallel format that is used in this amplifier.

It is interesting to observe that the bypassing and decoupling circuitry is as involved as the actual amplifier chain and its impedance-matching networks. This is a commonplace situation at VHF and UHF frequencies; it is necessary to discourage oscillation at lower frequencies, where the power gain of the stages greatly exceeds that which is available at the operational frequency.

Such an amplifier, with integrally contained driver stages, is relatively easy to implement because only 200 mW of input power is needed. Whereas the three cascaded stages operate "straight through" at 156 MHz, it is probable that one or more frequency multipliers will be used to process the excitation power. Such frequency-multiplying stages are useful in FM transmitters because they also multiply the FM deviation produced in a frequency-modulated oscillator or early buffer stage.

70-W 2-M broadband linear amplifier

The circuit in Fig. 7-2 is that of the Mirage/KLM PA 10-70BL linear amplifier. Microstripline techniques are used for the input- and output-matching networks. Additionally, the Motorola MRF245 RF power transistor has an internal low Q L section in its base lead. The overall effect of such matching networks is that no tuning is re-

70-W 2-M broadband linear amplifier

PA10-70 BL/PA 10-80BL
Parts list

Component	Value	Notes
C2, C3, C4	68 pF	Underwood
C5, C6	200 pF	Underwood
C8, C9, C16, C17, C22	0.001 μF	Disc
C7, C14	1 μF	35 Vdc Tant
C10, C11	100 pF	Underwood
C2, C12	47 pF	Underwood
C1, C13, C20, C21	750 pF	DM15
C19	5 pF	Disc
C23	25 μF	15 Vdc electrolytic
C18	47 μF	63 Vdc
C15	1000 pF	Feed through Allen Bradley
D3, D4	1N914	
R6	22 kΩ	1/4 W
R5	560Ω	1/2 W
RFC1, RFC3, RFC5	0.15 ΩH	
Q1	MRF245	Motorola
Q2	MPSA13	Motorola
R7	560Ω	1/4 W
R2	100Ω	Pot
R3	50Ω	1/4 W
R4	50Ω	5W
S1	Alco	SPDT
LED	MV5053	
R1	15Ω	1/2 W
RFC2	5T #16	1/4" dia. × 1/2" long
RY1	Relay 1355/1365	
D2, D1	1N4003 diodes	
F1	Fuse	15 A
RFC4	47 μH	

7-2 A 70-W, 2-m broadband linear amplifier. With a proper antenna and RF drive source, no tuning is required. KLM Electronics, Inc.

quired when the amplifier is used within the 143- to 149-MHz frequency range. The 70-W output rating applies for ICAS duty cycles. SSB, AM, FM, or CW operating modes can be accommodated. Input power can be in the 5- to 15-W range. The maximum input VSWR is 1.4:1.

The basic simplicity of the layout cannot fail to make an impact on those whose transmitter experience derives primarily from older equipment. Here, it is only necessary to connect the RF drive power, the dc operating power, and the antenna. The RF drive is intended to be provided from a transceiver.

Darlington transistor Q2 and its associated circuitry are a carrier-operated relay. With this provision, the contacts of relay RY1 activate the amplifier when the transceiver is placed in its transmit mode. This is accomplished by forward-biasing the RF transistor (it operates in class AB), delivering the RF drive to its base circuit, and connecting the antenna to its collector circuit. When thus enabled, the LED in the relay circuit turns on. Conversely, when the transceiver is placed in its receive mode, the contacts of relay RY1 disconnect the forward bias from the RF transistor and connect the antenna to the transceiver. Thus, once the operator has installed the amplifier, procedures remain substantially unaffected. When the transceiver is in its receive mode, the amplifier consumes only a negligible direct current from the battery (on the order of 10 mA or less). Thus, power switch S1 can be left on for long periods with no adverse effects.

Diode D1, mounted in close thermal proximity to the RF transistor, causes the forward bias to track the operating temperature so that the 10-mA collector "idling" current remains substantially constant over the operating temperature range of the amplifier. This not only prevents thermal runaway, but it ensures that the intermodulation distortion of the amplifier remains optimally low.

Performance curves for the Mirage/KLM PA 10-140BL amplifier are shown in Fig. 7-3. This amplifier is basically similar to the model PA 10-70BL, except for its greater output rating of 140 W. Notice how quickly the specified power capability can be lost with inadequate dc supply voltage and/or low RF drive. Although it is commonplace to refer to the automobile "12-V" storage battery, this does not correspond to a fully-charged battery. In practical installations of mobile equipment, the voltage-drop of inadequate wiring is often the culprit; even though the battery is generally maintained in a fully charged condition, the voltage actually applied to equipment (such as a linear amplifier) might be low enough to appreciably degrade performance.

80 W of class-C power

The uncomplicated amplifier in Fig. 7-4 conservatively develops 80 W at 30 MHz. The operational mode is class C, and input and output impedances are 50 Ω. This amplifier is eminently suited for CW operation. The power transistor is specially designed for high-frequency service up to 30 MHz. Motorola makes two versions that differ from each other in mounting provisions. The MRF454 has a mounting flange, whereas the MRF454A has a mounting stud. Both have opposed-emitter tabs to minimize emitter lead inductance and to diminish feedback from collector to base. The

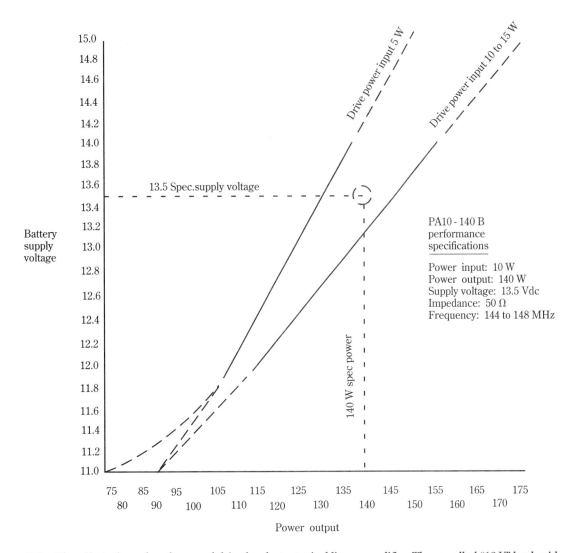

7-3 The effect of supply voltage and drive level on a typical linear amplifier. The so-called "12-V" lead-acid storage battery must be fully-charged (approx. 13.5 V) for optimum performance. Mirage/KLM Electronics

packaging technique greatly contributes to stability and is one of the reasons that no neutralization is needed in most implementations.

The circuit shown develops 12 dB of power gain and collector efficiency is 50%. Of course, without a dc negative bias applied to the base, the circuit does not operate deeply into the class-C region, as is commonplace with tubes. On the other hand, true class-B operation would require forward biasing to the threshold of emitter-base conduction. Because neither the transistor nor the circuit are expressly designed for linear amplifier service, the somewhat arbitrary class-C designation is useful to sug-

220 *Medium- and high-power applications*

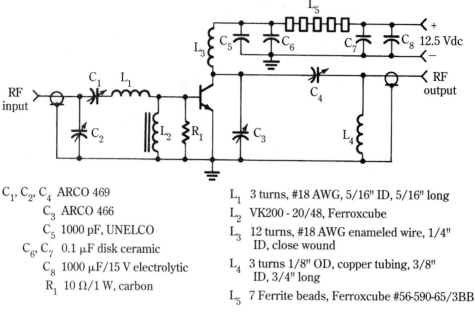

C_1, C_2, C_4 ARCO 469
C_3 ARCO 466
C_5 1000 pF, UNELCO
C_6, C_7 0.1 μF disk ceramic
C_8 1000 μF/15 V electrolytic
R_1 10 Ω/1 W, carbon

L_1 3 turns, #18 AWG, 5/16" ID, 5/16" long
L_2 VK200 - 20/48, Ferroxcube
L_3 12 turns, #18 AWG enameled wire, 1/4" ID, close wound
L_4 3 turns 1/8" OD, copper tubing, 3/8" ID, 3/4" long
L_5 7 Ferrite beads, Ferroxcube #56-590-65/3BB

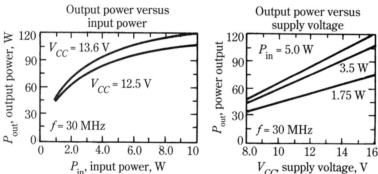

7-4 Conservatively rated 80-W power amplifier for 30 MHz. The operational mode and the intended service are suggestive of class-C tube amplifiers. _{Motorola Semiconductor Products, Inc.}

gest that the amplifier is not intended for SSB service. This is not to say, however, that the experimenter cannot coax acceptable linearity from the circuit by biasing the transistor into its class-AB region. However, the beautiful thing about modern solid-state RF practice is that the manufacturers make optimally suited transistors available for various operational modes. Thus, there are, in addition to "workhorses," specific transistor families for AM, FM, SSB, CW, and driver applications. Additionally, they are grouped in frequency ranges. Such service orientation makes solid-state RF practice more of a science and less of an art.

Incidentally, it is not wise to attempt to bias an RF transistor deeply into its class-C region, unless the transistor has been intended for such operation. Otherwise, such reverse biasing tends to make the transistor more vulnerable to destruction from secondary breakdown or hot spotting. That is why the base circuits of most amplifiers are designed to avoid the counterpart of self- or grid-leak biasing from the drive signal. For CW, keying is best done in a low-power driver stage.

The RF choke, L5, is composed of seven ferrite beads. A choke in this position must have the minimum inductance that is consistent with a reasonably high reactance at the operating frequency. This discourages the tendency of transistors to oscillate at low frequencies. For the same reason, the Q of RC choke L2 in the base circuit is "spoiled" by resistance R_1.

Mobile 80-W 175-MHz FM amplifier

The amplifying system in Fig. 7-5 incorporates unique features that are not commonly encountered when working with lower power levels. The first stage is biased for class-A operation, whereas the subsequent stages operate in class C. This enables the system to have a very low drive requirement; only 180 mW is needed. Because of the low power level at which the first stage operates, the class-A mode has minimal effect on overall efficiency, which is nearly 50%. The approximate power levels at the input and output of the four stages are shown in Fig. 7-6.

At first glance, the configuration of the output stage is suggestive of a push-pull circuit. Closer examination will reveal that the two transistors actually operate in parallel. This method of paralleling is superior to simply connecting together the respective terminals of the transistors. Although a little more involved, it largely circumvents or overcomes unequal load-sharing problems, as well as the problems that are associated with inordinately low impedance. The 10-Ω resistances in the base and collector circuits dissipate only that power, which results from imbalances in these circuits. In practice, such wasted power is small—even when the two transistors differ appreciably. Of course, if it is feasible to do so, it is preferable to select output transistors with closely matched characteristics. The 100-pF capacitors that are associated with the output tank are influential in attenuating second-harmonic energy; they are connected to the midpoints of each half of the C-shaped inductor. It is imperative that these capacitors have very low lead inductance. Special attention to second harmonic reduction is important because, unlike a push-pull circuit, the parallel connection displays no harmonic-canceling action.

Surprisingly clean output power is forthcoming from this amplifying system. The second harmonic is on the order of 38 dB down, and all other harmonics are greater than 50 dB down. The output stage dc current is 8.2 A when operating from a 12.5-V source. With the two 2N6084 transistors mutually mounted on a large heatsink, infinite VSWR under any phase angle will not destroy the transistors. Thus, in the extreme situations where the load inadvertently becomes either short- or open-circuited, no damage will ordinarily occur. Evidence of the inordinately high power gain of this amplifier (nominally 26 dB), can be seen in the performance curve of Fig. 7-7, where 180 mW of input power is sufficient for the rated 80-W output.

222 *Medium- and high-power applications*

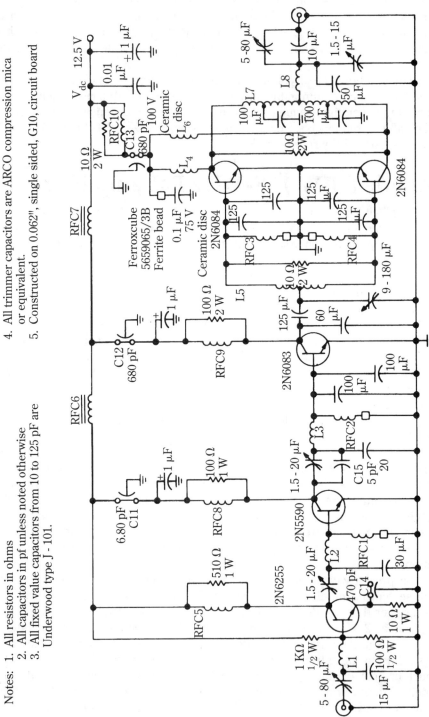

Notes: 1. All resistors in ohms
2. All capacitors in pf unless noted otherwise
3. All fixed value capacitors from 10 to 125 pF are Underwood type J-101.
4. All trimmer capacitors are ARCO compression mica or equivalent.
5. Constructed on 0.062", single sided, G10, circuit board

RFC1, RFC2, RFC3, RFC4	0.15 µH molded choke with Ferroxcube 5659065/3B ferrite bead on ground lead
RFC5	0.15 µH molded choke
RFC6, RFC7	Ferroxcube VK-200 19/4B ferrite choke
RFC8	4T #16 AWG wire, wound on 100-Ω 1-W resistor (75 nH)
RFC9	2T #15 AWG wire, wound on 100-Ω 2-W resistor (45 nH)
RFC10	10T #14 AWG wire wound on 10-Ω 2-W resistor
L1, L2, L3	1T #18 AWG, 1/4" dia, 3/4" L (25 nH)
L4, L6	2T #15 AWG wire, 1/4" dia, 1/2" L (30 nH)
L5, L7	See outline diagram.
L8	#12 AWG wire approximately 1" long (9 nH)
C11, C12, C13	680 pF, Allen Bradley type FA5C
C14	470 pF, Allen Bradley type SS5D
C15	5 pF, dipped silvered mica

Motorola Semiconductor Products, Inc.

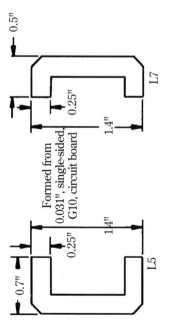

Outline diagrams for coils L5 and L7

7-5 An 80-W amplifying system for mobile 175-MHz FM service. Despite the resemblance of the output stage to a push-pull configuration, it is actually a unique form of a parallel circuit.

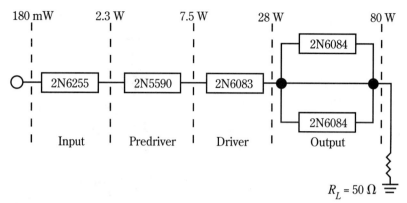

Power levels obtained using minimum specified device gains.

7-6 Power levels in the four-stage FM amplifying system. Notice that the two output transistors operate in parallel. Motorola Semiconductor Products, Inc.

Respectable power levels from MOSFET devices

The RF power amplifiers in Figs. 7-8 and 7-9 depict the vast strides that tend to characterize electronic technology in 10-year increments. A short time ago 100 W of VHF power from a MOSFET device was certainly a blue-sky concept. Yet these two amplifiers are now practical realities, and the MOSFET devices are designed and packaged for such high-frequency service.

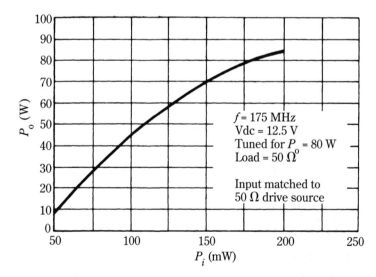

7-7 Output vs. input characteristics of an 80-W, 175-MHz amplifier. Intended for mobile FM service, this class-C amplifier operates from a nominal 12.5-Vdc source.
Motorola Semiconductor Products, Inc.

Respectable power levels from MOSFET devices

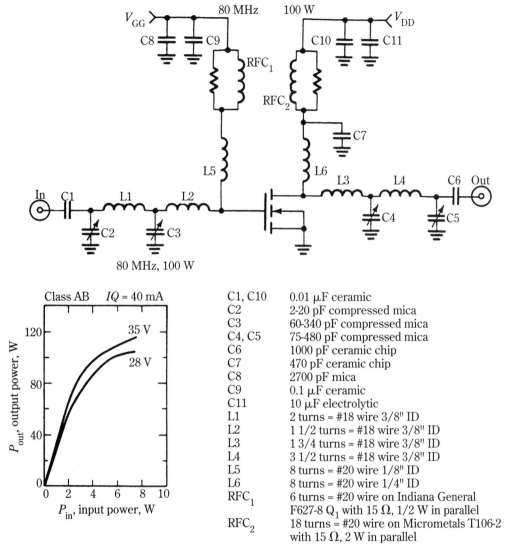

7-8 A 100-W, 80-MHz power MOSFET amplifier. Power MOSFET devices are now available in previously unthinkable power-frequency capabilities. _{Communications Transistor Corporation}

Although power MOSFETs are often considered to have infinite input impedance, such a notion is valid only for dc and low-frequency applications. In the RF range, the effective input impedance is low and continues to decrease with increasing frequency. It is, however, considerably higher than for a bipolar power transistor of similar output capability. This makes the MOSFET circuit easier to translate into practical hardware—especially where the combination of high power and high frequency is involved. Notice the difference in required input power for a given output level at 80 and 175 MHz.

226 Medium- and high-power applications

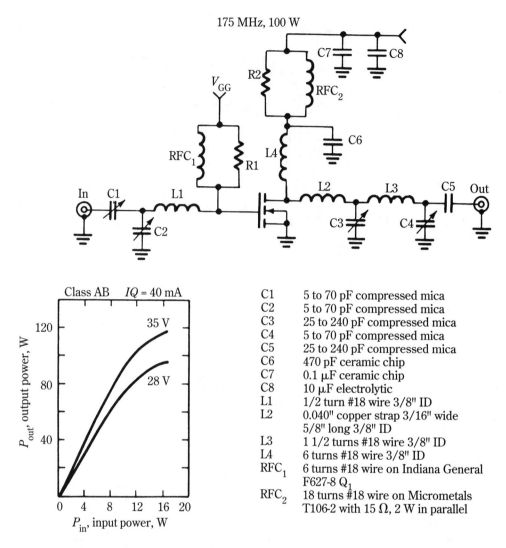

C1	5 to 70 pF compressed mica
C2	5 to 70 pF compressed mica
C3	25 to 240 pF compressed mica
C4	5 to 70 pF compressed mica
C5	25 to 240 pF compressed mica
C6	470 pF ceramic chip
C7	0.1 µF ceramic chip
C8	10 µF electrolytic
L1	1/2 turn #18 wire 3/8" ID
L2	0.040" copper strap 3/16" wide 5/8" long 3/8" ID
L3	1 1/2 turns #18 wire 3/8" ID
L4	6 turns #18 wire 3/8" ID
RFC_1	6 turns #18 wire on Indiana General F627-8 Q_1
RFC_2	18 turns #18 wire on Micrometals T106-2 with 15 Ω, 2 W in parallel

7-9 A 100-W, 175-MHz power MOSFET amplifier. Essentially a higher-frequency adaptation of the circuit of Fig. 7-8; similar output power is attainable.

The input and output impedance levels of both circuits are 50 Ω. Sufficient positive dc voltage must be applied to the V_{GG} terminal to cause a quiescent drain current of 40 mA (with no RF input). This adjusts the operating mode for class AB. Thereafter, circuit performance is straightforward, and the MOSFET is relatively immune to damage from high VSWR. These devices are tested to withstand infinite VSWR at all phase angles at rated output power of 100 W at 175 MHz and 28-Vdc supply.

The BF 100-35 RF power MOSFET used in these circuits is fabricated with beryllium oxide ceramics. This provides a low "junction"-to-case thermal resistance on the order of 0.7°C/W, and at the same time incurs negligible power dissipation. Remember that any mechanical or chemical treatment of this material is hazardous—the inhalation of even minute quantities of its dust or fumes can be extremely injurious to health. In ordinary handling procedures (where you pay due respect to the physical integrity of the device), there should be no danger.

Implementation of the balanced transistor

The 125-W broadband (225- to 400-MHz) amplifier in Fig. 7-10 uses a balanced transistor. This unique device actually contains two RF power transistors within a common package. The overall circuit arrangement is a push-pull configuration. Those familiar with push-pull amplifiers at lower frequencies or vacuum tubes are familiar with the second-harmonic canceling properties of such circuitry. Often overlooked, however, are two additional features that are of prime importance in the processing of high frequencies and high powers in transistor amplifiers.

Much of the art and science involved in the practical implementation of transistor RF power circuits involves the very low input and output impedances encountered. This makes it difficult to design matching networks with practical-valued elements. The installation and wiring of such networks often introduces stray parameters that are comparable in size to the network elements themselves. A second manifestation of this basic problem is in establishing a near-zero impedance ground for the emitter (or base in common-base circuits). The penalty for a tiny fraction of an ohm of resistance and/or inductance can be severely reduced gain and various instabilities. The balanced transistor and its accompanying push-pull circuit action help overcome these problems. The reason that such an approach had not previously been popular is that practical considerations render it difficult to implement for the combination of high frequency and high power. The two devices must be well balanced and the RF circuitry must be compact, with vanishingly short lead lengths.

In a balanced push-pull configuration, there is no need to provide a good RF ground for the mutual emitter terminals (or mutual base terminals in a common-base arrangement). The circuit, itself, establishes a "virtual" RF ground at its "center of gravity." This virtual RF ground is more effective than you could derive from physical grounding. Thus, the actual grounding of the mutual emitters is for the benefit of the dc, not the RF, path. For a given power level, input and output impedances of push-pull circuits are four times higher than for single-ended amplifier circuits. This greatly relaxes the design limitations of the matching networks. The fact that the shunt elements of these networks do not have to be grounded also helps practical implementation. Other means than the one illustrated can be used to accomplish transformation from unbalanced to balanced circuitry, and vice versa. For example, a simple center-tapped ferrite transformer merits consideration if somewhat less emphasis is placed on broadband response. As might be inferred, Communications Transistor Corporation also makes common-base versions of this dual device.

150-W 28-MHz test amplifier

Manufacturer's test circuits provide useful information for designers and experimenters. These circuits prove the basic performance of the device involved and are

228   Medium- and high-power applications

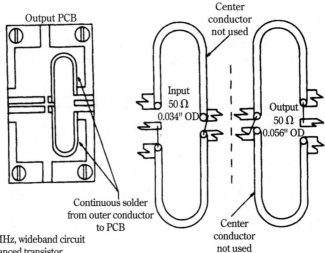

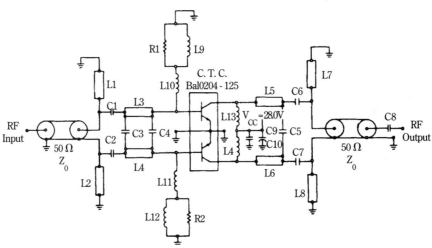

Schematic of 125 - W, 225 to 400 - MHz, wideband circuit using C T C Bal0204 - 125 balanced transistor

Broadband 225 to 400 - MHz amplifier

Bal0204 - 125	CTC balanced transistor for 225 - to 400 - MHz operation .
C1, C2	39 - pF ceramic - chip capacitor
C3	33 - pF ceramic - chip capacitor
C4	56 - pF ceramic - chip capacitor
C5	18 - pF ceramic - chip capacitor
C6, C7, C8	27 - pF ceramic - chip capacitor
C9	0.01 - µF ceramic capacitor
C10	10 - µF electrolytic capacitor
L1, L2, L3, L4, L5, L6, L7, L8	Printed on the circuit board
L9, L12	4.7 µH RF choke
L10, L11, L13, L14	0.1 µH RF choke
R1, R2	10 Ω 1/4 W

7-10   A typical implementation of the balanced transistor. Essentially two RF power transistors within a common package, this device improves the two main difficulties associated with transistor RF power circuits: inordinately low impedances and critical emitter lead length.   Communications Transistor Corp.

often implementable "as is," or nearly so. The high-power amplifier in Fig. 7-11 can be operated either as a linear amplifier or in class-C mode. The PT9790 RF transistor is electrically rugged; it is designed to withstand infinite VSWR and rated for 300-W dissipation at 25°C. It can perform well in SSB, FM, and AM modulation formats. The 50-V operating voltage results in an output impedance that is approximately 16 times that of a hypothetical like-powered device operating from a nominal 12.5-V supply. This is an important factor in the practicality of the amplifier circuit; an efficient output network for a 12.5-V device would not be easy to implement.

The input and output impedance-matching networks are similar, with L2 and L3 serving as RF chokes. The two chokes have different inductances to discourage self-oscillation. Inductor L1 is silver plated to maximize its Q. Extraordinary precautions

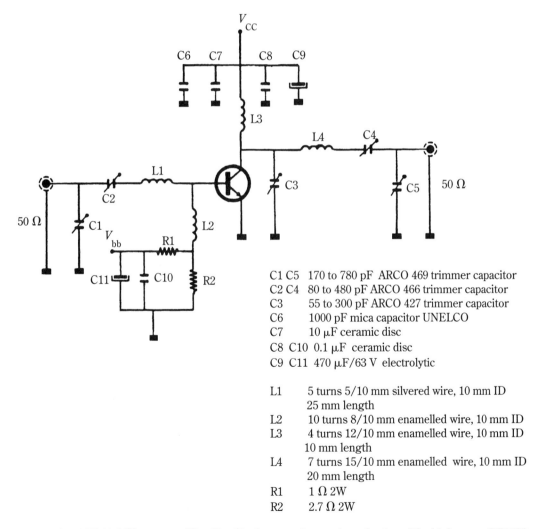

7-11 A 150-W, 28-MHz test amplifier. Used in the manufacturer's evaluation of the high-power PT9790 transistor, this straightforward circuit has versatile application possibilities   TRW RF Semiconductors

230 Medium- and high-power applications

are evident in the bypassing of the power supply. The objective is to prevent oscillation or instability at low frequencies, where the power gain is relatively high. The quality of the electrolytic capacitor, C9, assumes a vital role in this regard; one with a high ESR (effective series resistance) might not be adequate for the purpose.

A V_{bb} source that is sufficient to produce a quiescent collector current of 50 mA sets this amplifier up for class-AB linear operation. With no such base voltage applied, the amplifier operates in class-C mode. Special consideration is needed for the V_{bb} source. Ideally, it should track the thermal characteristics of the transistor in order to maintain the 50-mA quiescent current under actual operating conditions. An essentially heavier duty version of the PT9790 transistor is the LOT-1000, which can develop output power on the order of 200 W.

1 kW from 2 to 30 MHz

Throughout the years, the power rating of 1 kW has been a target for amateur radio transmitters. Even though the technical and legal implications are different for the various modulation formats, the construction of a solid-state amplifier with 1-kW output capability is an inherently useful project, where the mere novelty of solid-state power does not satisfy performance objectives.

This 1-kW amplifier is convenient and economic to build, and its operation is straightforward for the following reasons:

- The amplifier is a system that consists of four identical 300-W push-pull amplifiers.
- Output power from the four amplifiers is summed by a simple hybrid transformer arrangement. This scheme, at the same time, electrically isolates the four amplifiers from one another.
- 50-V transistors are used, thereby making impedance levels practically suited for readily constructed matching transformers.
- Because of the incorporation of the driver amplifier, the 1-kW system has an overall power gain of about 34 dB, making excitation requirements as benign as with many tube "finals." Moreover, the layout and construction of the driver are very similar to that of the four 300-W amplifiers.
- Stable class-AB operation prevails because of a thermally tracked bias supply.

The main performance characteristics of the system are shown in Figs. 7-12, 7-13, and 7-14. The constructional features of the power amplifiers are revealed in the photographs of Figs. 7-15 and 7-16. Notice that pairs of 300-W amplifiers are mounted on dual heatsinks—each with its own blower.

The block diagram of the 1000-W amplifying system is in Fig. 7-17. The format is essentially that of Fig. 1-9. Although the output can approach 1200 W, generally some loss is in the combining process and additional loss is in the low-pass filter, which should be interposed between the output and the antenna. Because of the push-pull circuitry and good linearity, the imposition made on such a filter is not great. All things considered, the deployment of the four 300-W contributing amplifiers enables a conservative 1000 W to be available at the antenna feeder line.

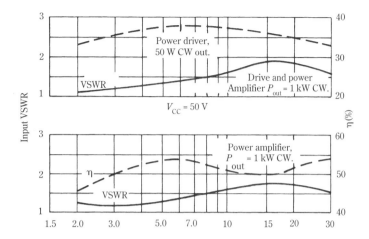

7-12 VSWR and efficiency versus frequency. The nominal power gain of the overall amplifier is 30 dB and is flat over the frequency range to within plus or minus 1.5 dB.
Motorola Semiconductor Products, Inc.

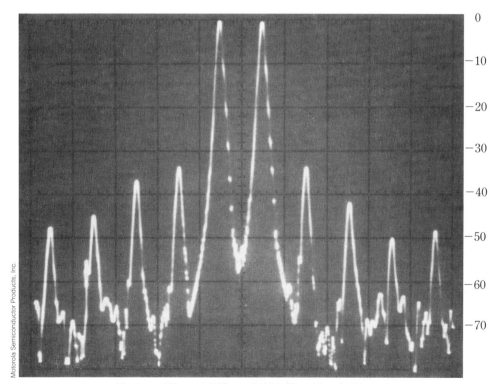

Test conditions: 1 kW out @ 30 MHz with a 50 V supply

7-13 A spectrum analyzer display of IMD products of the 1-kW RF System.

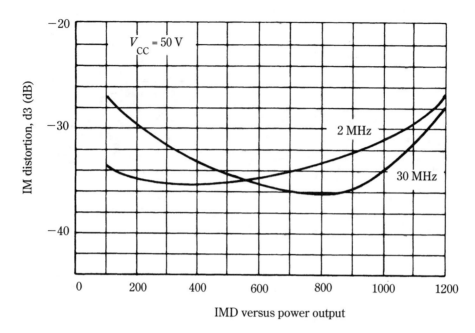

7-14 IMD versus power output. The third-order product, generally considered to be the most offensive, is depicted in these curves. Motorola Semiconductor Products, Inc.

The schematic diagram of the 300-W amplifier is in Fig. 7-18. As already pointed out, four such identical amplifiers are required in the system. Figure 7-18 is also the schematic diagram of the single driver amplifier. Although the circuitry of the 300-W amplifiers and the driver amplifier are the same, many of the components in the two amplifiers have different values. This, of course, is to be expected because of the disparity in power levels.

Each of the four 300-W amplifiers and the driver amplifier has its own thermal-tracking bias circuit. The schematic of the bias source for both amplifiers is in Fig. 7-19.

The parts list for the 300-W amplifiers and for the driver amplifier is shown in Table 7-1. The RF power transistors, Q1 and Q2, are not identified in the parts list. These transistors are the Motorola MRF428 for the 300-W amplifiers and the Motorola MRF427 for the 50-W driver amplifier.

The semipictorial presentation of the output transformer in Fig. 7-18 makes clear the balun nature of this transformer. Do not construe, however, that the single-turn "windings" apply. Rather, the actual number of turns is delineated in Table 7-1. Also, the construction of the output transformer for the driver amplifier differs from that for the 300-W amplifiers. This information, also, is given in the parts list.

Notice in Fig. 7-18 that the secondary of input transformer T1 is not center-tapped. This is permissible because base-current return paths occur through the forward-biased base-emitter junction of the "off" transistor. Another interesting aspect of this push-pull circuit is the use of transformer T2 in place of separate RF chokes. This scheme enables the requisite reactance to be developed with relatively low I^2R loss, an important matter at high power levels. It also allows a simple means to provide negative feedback via ter-

Table 7-1. Parts list for 300-W power amplifiers and for 50-W driver amplifier. The circuit and parts layout are the same for both amplifier boards. The 300-W power amplifiers use a pair of MRF428 RF power transistors. The 50-W driver amplifier uses a pair of MRF427 RF power transistors.

	Power module	Driver amplifier
C1, C2	5600 pF	3300 pF
C3	56 pF	39 pF
C4	470 pF	Not used
C5	560 pF	470 pF
C6	75 pF	51 pF
C7, C8	0.1 µF	0.1 µF
C9, C10	0.33 µF	0.33 µF
C11	10 µF/150 V	10 µF/150 V
R1, R2	2 × 3.9 Ω/½ W in parallel	2 × 7.5 Ω/½ W in parallel
R3, R4	2 × 6.8 Ω/½ W in parallel	2 × 18 Ω/½ W in parallel
L1, L2	Ferroxcube VK 200 19/4B ferrite choke	Ferroxcube VK 200 19/4B ferrite choke
L3, L4	6 ferrite beads each, Ferroxcube 56 590 65/3B	6 ferrite beads each, Ferroxcube 56 590 65/3B

All capacitors, except C11, are ceramic chips. Values over 100 pF are Union Carbide type 1225 or 1813 or Varadyne size 18 or 14. Others ATC Type B.

T1	9:1 type	4:1 type
	(Ferrite cores for both: Stackpole 57-1845-24B or Fair-Rite Products 287300201 or equivalent.)	
T2	7 turns of bifilar or loosely twisted wires. (AWG #20.) Ferrite cores for both: Stackpole 57-9322, Indiana General F627-8Q1 or equivalent.	
T3	14 turns of Microdot 260-4118-00* 25 Ω miniature coaxial cable wound on each toroid. (Stackpole 57-9074, Indiana General F624-19Q1 or equivalent.)	11 turns of RG-196, 50 Ω miniature coaxial cable wound on a bobbin of a Ferroxcube 2616P-A100-4C4 pot core.

*Equivalent product is available from: W. L. Gore, Inc., Newark, Del. Part number CXN1286

ªMotorola Semiconductor Products, Inc.

tiary winding L5. Decoupling and bypassing of the 50-V supply are accomplished with two separate paths that involve L3 and L4, rather than by means of a mutual path, as is conventionally done. This enhances the low-frequency stability of the amplifier.

Thermal-tracking bias source

The circuit of the thermal-tracking bias source in Fig. 7-19 is essentially a voltage regulator with a current-boosting transistor, Q3. A significant modification from conventional voltage regulators consists of the inclusion of "diode" D1 in the internal voltage-reference circuit of the MC 1723 IC regulator. D1 is actually the base-emitter junction of a plastic-packaged 2N5190 transistor. D1 is mounted near the center of the amplifier board, close to the push-pull RF power transistors, Q1 and Q2.

234 *Medium- and high-power applications*

7-15 A pair of 300-W amplifiers, such as Units A and D in the block diagram. The power splitter is in the foreground.

The voltage developed across D1 decreases with rising temperature, as does the required bias voltage of the RF power transistors. As the temperature increases, the responding junction voltage of D1 causes the regulator to reduce its output voltage, which thereby keeps the quiescent current of the amplifier approximately constant. Such near-constancy is necessary to prevent thermal runaway and is also important to maintain low intermodulation distortion. For the latter situation, it is necessary for the bias source to have a low dynamic impedance; this is the reason why the voltage-regulating circuit is used. Actually, the bias control is indirect; there is no closed loop in which the actual amplifier current is sensed. The reason this scheme tracks in a satisfactory manner is that the base-emitter voltages of D1 and of the power transistors are governed by virtually the same temperature function.

The nominal output voltage of the bias source is adjustable by means of potentiometer R10. For the 300-W amplifiers, the 25°C quiescent current is adjusted by R10 to 300 mA. For the driver amplifier, the adjustment is for 80 mA. At a heatsink temperature of 60°C, these values will hold sufficiently so that the oper-

7-16 A pair of 300-W amplifiers, such as Units B and C in the block diagram. Power combiner is in the foreground.

ating points of the RF power transistors remain substantially at their optimal class-AB values. Actually, the amplifier current decreases somewhat at the higher operating temperatures—evidence of a small amount of overcontrol. This appears to be entirely satisfactory and reinforces the conviction that there is no gradual "creep" toward thermal runaway.

236 Medium- and high-power applications

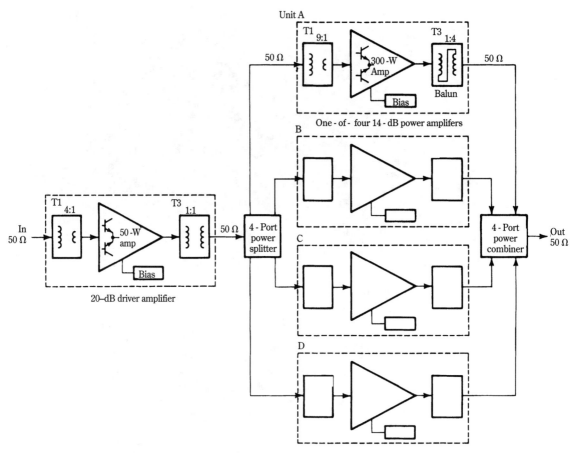

7-17 A block diagram of 1-kW, 2- to 30-MHz Amplifying System. The dashed lines indicate complete circuit boards.

Input power splitter

To avoid confusion, remember that all that has been described pertains to the four 300-W power amplifiers and to the 50-W driver amplifier. All subsequent discussion deals with the technique of summing up the power from the four 300-W amplifiers so that a conservative kilowatt will be available for the antenna feeder line or other 50-Ω load. The power-summing process is accomplished with the aid of a power splitter (divider) and a power combiner. The former device splits the power from the driver amplifier into four equal parts for application to the inputs of the four 300-W amplifiers. The power combiner channels the output power from these four amplifiers into the load. Both the splitter and the combiner also isolate the four amplifiers from one another. A practical manifestation of this is that, if one amplifier becomes inoperative, the remaining three amplifiers will continue to operate at undisturbed impedance levels and the load will still receive about 750 W.

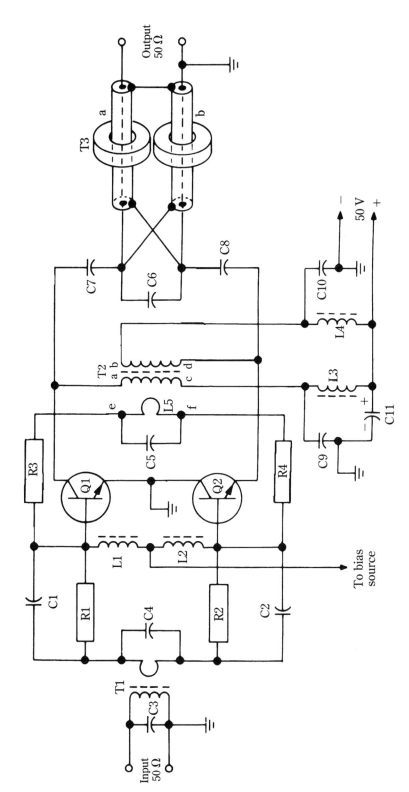

7-18 The circuits of the 300-W power amplifier and of 50-W driver amplifier. Substantially constant power gain from 2 to 30 MHz is developed by mutual action of negative-feedback network L5, R3, R4, and of input networks R1, C1, and R2, C2. *Motorola Semiconductor Products, Inc.*

238 Medium- and high-power applications

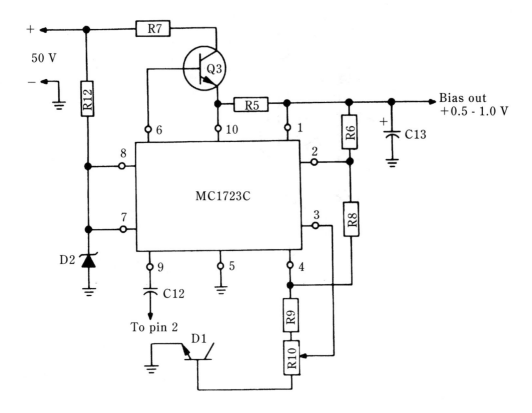

Parts list:
R5 1.0 Ω 1/2 W
R6 1 kΩ 1/2 W
R7 100 Ω 5W
R8 18 kΩ 1/2 W
R9 8.2 kΩ 1/2 W
R10 1 kΩ trimpot

R12 1 kΩ 1/2 W
C13 1000 μF 3V electrolytic
C12 1000pF ceramic
D1 2N5190
D2 1N5361 1N5366
Q3 2N5991

7-19 A thermal-tracking bias source for the 300-W power amplifiers and for the driver amplifier. Each amplifier board incorporates one of these bias sources. D1, the sensing diode, is the base-emitter junction of a plastic-packaged 2N5190 transistor.
_{Motorola Semiconductor Products, Inc.}

Both the power splitter and the power combiner use hybrid transformer circuits in which the transformers are transmission-line balun types. Additionally, both the splitter and combiner use an additional transformer for impedance matching.

Figure 7-20 is the schematic diagram of the power splitter. Depicted are four ferrite-bead loaded coaxial lines, an impedance-matching transformer, and four balancing resistors. The pictorial representation of the coaxial baluns is realistic; they consist of ferrite beads over a 1.2-inch length of RG-196 coaxial cable. The ferrite beads are Stackpole 571511-24B, Fair-Rite Products 2673000801, or an equivalent.

The impedance-matching transformer is seven turns of 25-Ω miniature coaxial cable, such as Microdot 260-4118-000, on a ferrite toroid. The toroid is a Stackpole

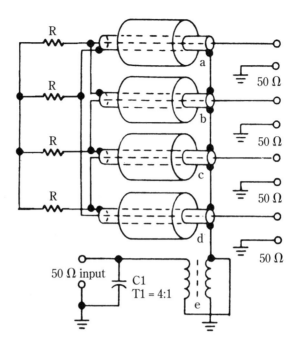

7-20 A 4-port input splitter. The balun transmission-line elements are, as depicted, lengths of coaxial cable passed through ferrite sleeves. _{Motorola Semiconductor Products, Inc.}

57-9322-11 or an equivalent. This transformer is actually a transmission-line balun and the two "windings" are the inner conductor and the outer sheath of the coaxial cable. The 50-Ω impedance level from the driver amplifier is stepped down to the 12.5-Ω input impedance of the four-balun power splitter. Capacitor C1 might or might not be needed; its value can be empirically determined in terms of satisfactory input VSWR over the 2- to 30-MHz range.

The balancing resistors, R, dissipate negligible power during normal operation of the system. In the event of failure in one or more of the 300-W power amplifiers, these resistors will dissipate varying amounts of power. If these four resistors are each 28.13 Ω, the failure of one 300-W amplifier will have little effect on overall system operation, other than a reduction of power delivered to the load. The driver amplifier and the remaining three power amplifiers will otherwise operate under similar conditions to those that prevail during normal operation.

If it is OK to have such operational redundancy for the failure of two 300-W power amplifiers, the proper value of these balancing resistors is 25 Ω. The prospect of such a double failure might be construed a reasonable probability because of the physical arrangement of the four power amplifiers; they are constructed as two pairs of 600-W power modules.

In any event, the balancing resistors should all be equal in value and should be noninductive types. For most purposes 10-W power ratings should be sufficient. Possibly, resistance values in the neighborhood of 26.5 Ω or so could provide a compromise situation for failure of either one or two of the 300-W power amplifiers. Under

240 *Medium- and high-power applications*

this condition, the VSWR, linearity, and efficiency of the remaining amplifiers would be somewhat affected, but perhaps within acceptable limits.

Output power combiner

The output power combiner operates on similar principles to that of the input power splitter. Power flow is in the opposite direction because the objective is now to channel four contributive sources of power into the load. With the higher power levels involved, differences naturally occur in the construction of the elements.

Figure 7-21 is the schematic diagram of the power combiner. As with the power splitter, it has four ferrite-sleeve loaded coaxial lines, an impedance-matching transformer, and four balancing resistors. The pictorial representation of the coaxial baluns is again realistic; they consist of ferrite sleeves over a length of RG-142B/U coaxial cable. Each ferrite sleeve is a Stackpole 57-0572-27A.

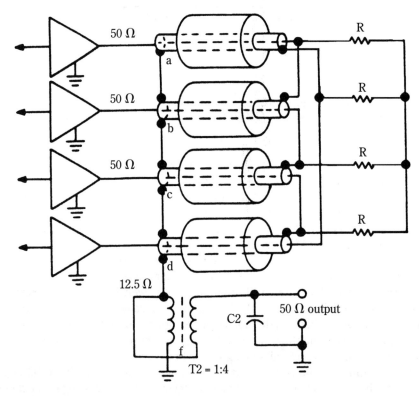

7-21 A 4-port output combiner. The balun transmission-line elements are, as depicted, lengths of coaxial cable passed through ferrite beads. Motorola Semiconductor Products, Inc.

8
Standing-wave ratio and related controversies

ONE OF THE MAIN FACTORS THAT IS RESPONSIBLE FOR THE WIDESPREAD confusion concerning the interpretation of standing-wave ratio (SWR) is the imprecise terminology and the vague semantics often appearing in technical literature. Then too, there is no shortage of false notions which, unfortunately, could not be fixed by even the most rigorous use of the language.

First examine the words that are commonly used to describe transmission-line and antenna behavior, together with other factors involved in convoying RF power from source to load. After all, it is widely held that all that needs to be done is to tune for a low SWR. Having obtained this objective, "common-sense" seems to say that the radiating efficiency of the antenna load is necessarily optimized, particularly if the SWR is very close to 1.0. This is not necessarily so. But first, let's play with some of the much-used terms.

To begin with, *SWR*, to be sure has entered the language; it abbreviates *standing-wave ratio*, but the technically literate know that what is usually meant is *voltage-standing-wave ratio (VSWR)*. This distinguished it from *power standing-wave ratio (PSWR)*, and from *current standing-wave ratio (CSWR)*. Although bridges, directional couplers, and reflectometers commonly provide readouts that are labeled *SWR*, this book will henceforth use VSWR for the sake of mathematical exactitude. I have observed mixups by otherwise astute technical writers because of failure to clearly distinguish between voltage and power standing-wave ratio.

Surprisingly, more than one variety of resistance exists in order to pin down unique aspects of certain otherwise "resistive" parameters. Consider, first "ordinary," ohmic resistance. There is generally no problem here because of its universally known capacity of dissipating electrical power. The dissipation is most often in the nature of heat, but other energy can occur (such as light, acoustics, or mechanical displacement). An incandescent light bulb, by virtue of its filamentary resistance, produces some light, a lot of heat, and some other effects (such as metallic expansion). However, an electric motor cannot be said to develop mechanical displacement because of its resistance. For all practical purposes, it is safe to simply say that ordinary resistance converts electrical energy into heat energy. Table 8-1 describes commonly encountered resistive entities.

Table 8-1. The various types of resistances and resistance-like entities.

Type of resistance	Nature	Where encountered
Ordinary or positive	Dissipates electrical energy in form of heat. Vanishes near absolute zero for certain materials.	Heaters, resistors, and inadvertently in conductors and devices that carry current.
Negative resistance	Cancels positive resistance when associated with other devices. Supplies, rather than absorbs, energy in oscillators.	Directly in dynatron and transitron oscillators; indirectly in all LC oscillators, tunnel diodes, and Gunn diodes.
Characteristic impedance	A resistive value such that when used as termination on a transmission line, no reflections or standing waves occur.	RF transmission lines. Also has the same significance for LC filters where optimum performance is obtained when this value is used for the terminations.
Nonlinear resistance	Current change is not proportional to voltage change.	Lamp filaments, diodes, etc.
Radiation resistance	Acts resistively as termination for a transmission line, but it is nondissipative. Load effect of the universe.	Antennas
Dynamic resistance	Resistance is presented by the device for very small excursions of voltage and current. The resistance seen by a very small signal.	Characteristic curves of active devices, such as tubes and transistors.
Dc Resistance	Resistance offered to direct current.	Dc and low-frequency ac circuits.
Ac resistance	Resistance offered to ac where skin-effect tends to concentrate current at surface of a conductor.	In RF tanks and circuits and in transmission lines and antennas.

If ordinary resistance is designated as *positive resistance*, there is also *negative resistance*, which cancels or removes the positive resistance in a circuit. Thus, a number of negative-resistance devices, when associated with an LC tank circuit, cause oscillation to occur. This happens because such devices effectively remove the positive resistances in the resonant circuit, which dampens any tendency toward self-oscillation. Interestingly, nearly the same effect can be observed by cooling an LC circuit to a near-zero temperature. Although the mechanism is different in the two processes, both remove ordinary, or positive resistance that is normally inherent in any practical LC circuit at room temperature. Negative-resistance devices that have been, and are used, are the dynatron oscillator (using a tetrode electron tube), arc transmitters, the tunnel diode, and the Gunn diode. Also, any feedback oscillator in a "black box" can be analyzed mathematically as if it was a negative-resistance oscillator. Whereas ordinary resistance dissipates electrical energy, negative resistance effectively neutralizes the tendency for such energy loss.

Resistance of either variety can also be nonlinear. This implies a variation of resistance as a function of voltage or current. Thus, a tungsten lamp is a nonlinear resistance because its resistance increases with current. A carbon-filament lamp is also a nonlinear resistance, but in this case, the resistance decreases with current. Incidentally, dynamic resistance designates the resistance over a small portion of the voltage-vs-current relationship of a nonlinear device.

Yet another type of resistance is the *radiation resistance* of an antenna. This resistance is nondissipative in the usual sense; rather than the production of heat, a desirable phenomenon occurs: radiation. Yet, as the electromagnetic waves escape into space, they impose a loading effect on the feedline and on the transmitter that simulates that of an ordinary resistance. Indeed, the feedline and the transmitter cannot distinguish between a resonant antenna, which radiates, and a "dummy" load (a physical resistance), which does not. At this point, notice that a dummy load, with its possible SWR of 1.0, has a 0 radiation efficiency. Thus, the achievement of a near-unity VSWR when tuning up an antenna system, is contrary to widely held notions—no guarantee of efficient RF radiation from the antenna. Instead, the low VSWR measured at the transmitter end to the feedline could just as well be showing that the feedline, and perhaps the ground system are unusually lossy and that the transmitter "thinks" it is operating into a "dummy" load!

Here is a more practical aspect of radiation resistance (and radiation efficiency) than is derived from discussions of "coupling to the universe" or from speculations of the effect of cryogenic temperatures. It is practically feasible to operate an antenna considerably off its self-resonant frequency with negligible penalty in the efficiency of the antenna system, or in any meaningful change in the radiation resistance of the antenna. Just forego the usually sought goal of a 1.0 VSWR at the sending end of the transmission line. This does not incur high losses in typical amateur radio installations. An antenna that is self-resonant in the middle of the 80-M band will operate at either band extremity with very nearly the same efficiency as at its self-resonant frequency. To operate the antenna off-resonance, simply restore resonance to the antenna system, which is the combination of the antenna and feedline. The convenient way to do this is to insert an impedance-matching network or transmatch between the transmitter and the sending-end of the feedline.

Before the widespread use of coaxial cable the situation was well understood and low-loss open-wire feedlines were operated at VSWRs of 10 or 20 with no detrimental results. Of course, VSWR indicators weren't available to confuse the issue!

This is not to say that operating with the lowest possible VSWR is not a worthy objective. At frequencies higher than 30 MHz, and with long feedlines, the loss penalty for VSWRs greater than two certainly merits consideration.

This is an appropriate point in the concern over basic definitions to deal with several much used, and often abused, terms that pertain to transmission lines. Table 8-2 briefly delineates some descriptions that do not always enjoy consistent meanings in the technical literature. As already mentioned, the ongoing controversies about the interpretation and significance of VSWR stems greatly from nebulous use of the language. When a radio amateur laments the high VSWR on his transmission line, it often is not clear whether the reference is to the VSWR indication at the junction of the transmitter and the transmatch, the VSWR at the sending end of the feedline, or at the load-end of the feedline. Depending on a more precise reference of the allegedly high VSWR, a VSWR over 2.0 might be undesirable, or, conversely, that a VSWR as high as 10 or 15 will not endanger or degrade operation.

The radiation resistance of antennas conventionally pertains to the apparent resistance at a current loop on the antenna after any reactance is neutralized. Neutralization of antenna reactance can stem from self-resonance or from impedance matching at the feeder-antenna junction. Radiation resistance can be thought of as the measure of the degree that the antenna is coupled to the universe. From this viewpoint, the higher the radiation resistance, the greater should be the radiating efficiency of the antenna. However, in practice, this is not necessarily so. What is important is the ratio of radiation resistance to the resistance that is represented by the total dissipative losses in the antenna system.

You might suppose that a wire that is electrically much too short to form a self-resonant antenna, could, nevertheless become an efficient radiator if, by feeder impedance matching, such an antenna could be made to absorb sufficient RF current. It is always found, however, that when carried too far, the system losses become the controlling factor in trying to make the short antenna accept power from the transmitter. For example, a 20- or 40-M resonant antenna can be, by means of a base-loading inductance, made to resonate at 160 meters. The efficiency of such an antenna system would be very low because of the inevitable losses in the leading coil. By comparison, a longer antenna, even one with comparable radiation resistance, could be a more efficient radiator of RF energy if its overall dissipative losses were relatively low. The point is that radiation resistance alone does not guarantee efficient radiation; it is also necessary for the losses of the entire antenna system to be low.

Because dissipative losses waste power that could otherwise be radiated, it is not too far-fetched to ponder the effect of operating an antenna system at cryogenic temperatures, where ohmic losses tend to vanish. This, of course, would be neither practical nor economical for ordinary earth-bound use. However, space vehicles do operate in extremely cold environments. Fortunately, land-based space communications equipment does not need to be plagued by electrically small antennas. The preponderance of RF power fed into such antennas goes into radiation resistance. Only a small amount is consumed by "lossy" resistance.

Table 8-2. A brief description of transmission lines.

Transmission line terminology	Description or meaning
Feeder line	In this book, *feeder line* always refers to the transmission line that connects to the antenna or load. In typical installations, it is a relatively long line that transfers RF power from the transmatch to the antenna. It is not the short cable that connects the transmitter to the transmatch. Although the technical literature seems divided on this definition, it is very important to understand what line is being referred to when discussing VSWR or impedance levels.
Resonant line	This term usually implies a resonant condition as a result of the combination of the transmatch, the feeder line, and the antenna itself. Such resonance causes standing waves on the feeder line, but cancels reactance at the antenna feedpoint, and thereby enables the antenna to absorb RF power. In practice, it is common to refer to lines that are not flat, that is, lines with mismatched loads as "resonant lines". It usually pertains to VSWRs of 3.0 or higher.
Flat line	A transmission line that works into a matched load. Such a line displays a VSWR of 1.0 and does not show evidence of standing waves. Regardless of length, the sending-end impedance of the flat line is Z_0, the line's characteristic impedance.
Length of line	Although thought of in terms of physically measurable length, what really counts electrically is the length in wavelengths, and then only when corrected by the velocity factor.
Balanced line	Balanced line generally refers to open-wire, or twin-conductor transmission line in which equal-magnitude RF currents flow in the two conductors at any given line position. Such operation greatly reduces any tendency of the line to radiate.

All told, there is no transmitting equivalent of the extremely electrically short active receiving antenna, where a short whip is sufficient for very low frequencies. The active antenna uses an RF current amplifier with exceedingly high input impedance. Such an amplifier makes up for the tiny signal that the whip intercepts from space. Any attempt to simulate such operation for transmitting would be prohibitive in terms of wasted power and its attendant practical considerations.

The next type of resistance is not as easy to grasp without resorting to rather complex mathematics as the aforementioned resistive entities. Even worse, it is not entirely clear that rigorous mathematical treatment necessarily leads to useful practical insights. A case in turn involved early plans to lay a trans-Atlantic cable for telephone communications. Mathematicians argued between themselves regarding the feasibility of using such a long cable. The relevant mathematics was available, but quite different interpretations were drawn. Indeed, the correct interpretation by Oliver Heaviaide was initially rejected and his work had to be vindicated by Michael Pupin who took the practical step of using so-called loading coils to overcome the frequency distortion of long cables. The interesting aspect of this accomplishment is that it not only wasn't evident to mathematicians, but it also appeared to be counter to "common-sense" that asserted that adding series inductance could only cause more high-frequency attenuation and worsen the distortion of voice signals.

Fortunately, the counterparts of the submarine cable, coaxial and open-wire transmission lines, are favorable to practical manipulation without going off the deep-end mathematically. However, certain concepts must be properly mattered because their performance is not at all like a conventional "go and return" low-frequency systems (such as the wiring of a doorbell that operates from a 60-Hz transformer). Sometimes it is not easy to visualize a new or different way of relating cause-and-effect relationships.

The concept of surge impedance is otherwise known as the characteristic impedance of RF transmission lines, Z_o. To begin with, Z_o, for all practical purposes in RF work, can be treated as a pure resistance, R_o. That is, a more relevant designation of this parameter would be characteristic resistance. This can considerably simplify your understanding of this entity, which still requires a bit of mental effort to distinguish its nature from any of the resistance types, which were previously discussed. To keep matters as simple as possible, assume that you are dealing with lossless lines; those that have no equivalent series-resistance of the ordinary type, and no equivalent "leakage" conductance. That is, the lines are infinite numbers of to "pure" series inductances and infinite numbers of "pure" shunt capacitors.

It has been assumed that our transmission line is lossless; it is devoid of any ohmic-type series resistance and it has infinite shunt resistance. Although such a line carries RF power with no dissipation, it, will, nonetheless, behave as if it has a built-in resistance. This "lossless" resistance can be "seen" by both source and load. As mentioned, this resistance is by custom, referred to as an impedance, Z_o. The custom originated because it can be mathematically demonstrated that both, resistance and reactance can be "seen" under certain conditions. But, those conditions do not prevail in all ordinary uses of transmission lines at radio frequencies. So, for practical considerations, whether you deal with open-wire line or coaxial cable, or even microstrip, the characteristic impedance is actually resistive, At the same time, its resistive nature certainly differs from any of the previously discussed resistances. Table 8-3 shows relationships between Z_o and the physical constants of different types of transmission lines.

The best practical definition of characteristic impedance is: if an RF source feeds a transmission line of any length, there will be a terminating load resistance that will produce no reflected power that will provide the cherished VSWR goal of 1.0 at the sending end of the line. Any other value of load resistance, load reactance, or load impedance will reflect some of the power back to the sending end, where it will be re-reflected and will cause a VSWR reading other than the magic 1.0. A corollary of this statement is that the transmission line will be free of standing waves, or will be "flat" only when the line is terminated by a pure resistance, which is defined as the characteristic impedance of the line. Notice that no stipulations have been put on the

Table 8-3. The characteristic impedance of lines from physical dimensions.

Type of line	Characteristic impedance, Z_0 in Ohms	Parameters
Coaxial	$Z_0 = \dfrac{138}{\sqrt{k}} \log_{10}\left(\dfrac{D}{d}\right)$	k is the dielectric constant of the insulating material ($k = 1.0$ for air). D is the inner diameter of the outer cylinder. d is the outer diameter of the inner cylinder (D and d must be measured in the same units).
Open wire or twinlead (two conductors separated by insulating material, such as 300-Ω TV line).	$Z_0 = \dfrac{276}{\sqrt{k}} \log_{10}\left(\dfrac{S}{r}\right)$ A useful form of this formula for open-wire lines enables determination of spacing for a desired Z_0: $S = r \times 10^{\frac{Z_0}{276}}$ in which case S will be expressed in the units chosen for r	k is the dielectric constant of the insulating material ($k = 1.0$ for air). S is the center to center spacing of the wires. r is the radius of the wires (S and r must be measured in the same units).
Microstrip line	$Z_0 = \dfrac{377}{\sqrt{k}} \left(\dfrac{T}{W}\right)$	k is the dielectric constant of the board material. T is the conductor thickness. W is the conductor width (W and T must be measured in the same units).

driving impedance of the RF source. These statements are borne out in the standing-wave patterns shown as a function of load resistance in Fig. 8-1.

Because I like simplicity in practical applications, it is relevant to further state that even if you are dealing with real-life lossy lines, the value of the characteristic impedance is not influenced to any great extent by the ohmic resistances that make the line lossy. The characteristic impedance not only remains substantially intact, but it is not the agency of the dissipative losses in the line. The characteristic impedance is, at once, a mathematical entity with a nondissipative nature and a practical reality that manifests its presence (as has been described). It can be further pinned down as having a value equal to $\sqrt{L/C}$, where L is the series inductance of the line per unit length, and C is the shunt capacitance of the line per unit length. The concept of characteristic impedance has been kept quite simple in this book. It is nevertheless the all-important parameter that must be dealt with in RF systems. Fortunately, it is, in most instances, provided for us by the manufacturer of coaxial cables.

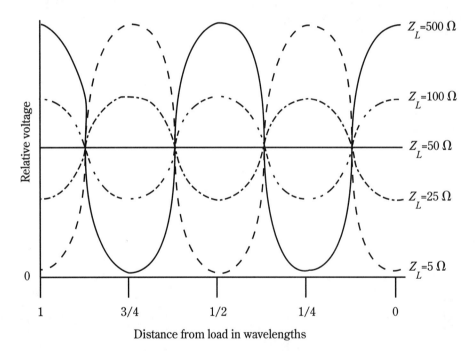

8-1 The effect of load resistance, Z_L, on a 50-Ω transmission line. For reactive loads, the same pattern prevails, but it is shifted to the right or left.

A quick review and a bit of amplification of an interesting mathematical relationship in transmission lines is: suppose an impedance bridge, vector voltmeter, or other instrument is connected to monitor the input end of a transmission line. Two impedance readings are then recorded. First, a reading is taken with the far end of the transmission line shorted. Then, a second reading is taken with the far end of the line open. These readings are simple magnitude data. Let the short-circuit reading equal Z_{sc}; let the open circuit reading equal Z_{op}. Then, the characteristic impedance of the transmission line:

$$Z_o = \sqrt{Z_{sc} \times Z_{op}}$$

Notice that this relationship can be construed as another definition of the characteristic impedance of transmission lines. Again, you are dealing with a practical entity that is defined by a mathematical relationship. As previously emphasized, Z_c is (for practical purposes at radio frequencies) resistive, but unlike "ohmic" resistance, it is disspationless.

In this instrumentation technique, the line can be of any length (in principle), and can be measured at any frequency. In practice, it is best for the line to not be too short electrically. A frequency of several megahertz is generally convenient to work with. Coaxial line is easy to measure because several or more tens of feet can be conveniently coiled up with no detrimental effect. Open-wire transmission line should not be coiled; but should be laid out straight. Unless inordinately high, the dissipative losses in the line do not materially affect the readings in this measurement technique. A particularly useful application of this method of determining characteristic impedance is the measurement of Z_0 for twisted lines that are often used in transmission-line transformers. It is perfectly acceptable to use a longer section of twisted line for the measurement than will be used in the actual transformer. Z_0 varies with the type and thickness of wire insulation, wire diameter, and tightness of twist.

An alternative, and often a more convenient alternative to the previous measurement technique is:

$$Z_o = \sqrt{\frac{L_{sc}}{C_{op}}}$$

where:

L_{sc} is the inductance in henries with the far end of the line shorted.
C_{op} is the capacitance in farads of the far end of the line open.
The same length of line must be used when measuring L_{sc} and C_{op}.

The electrical length of transmission lines

In general, length is an important physical dimension of transmission lines. Although VSWR will not change as the length of a lossless line is varied, the impedance "seen" at the sending end of such a line undergoes change, except with the notable exception of a flat line—a line properly matched at its load end so that no standing waves exist along the length of the line. In practice, the VSWR does, indeed, change with line length, but this is caused by the losses of nonideal lines. A confusing consequence of this effect of losses is that the VSWR at the sending end of the line improves if the length of the line is greater. This effect stems from overall attenuation in the line; if this effect is pronounced, it signifies that the line is too long, that the quality of the line is poor, or a combination of both situations. Remember that the ideal (lossless line) can exhibit variations in RF voltage, current, power, and impedance as a cyclic function of line length, but VSWR is constant over the length of such a line.

In practice, an efficient transmission-line system will behave much like the ideal situation and it is a mistake to try to improve VSWR by changing the line length. If VSWR is monitored at the load end of a practical line, you can confirm that only the

250 *Standing-wave ratio and related controversies*

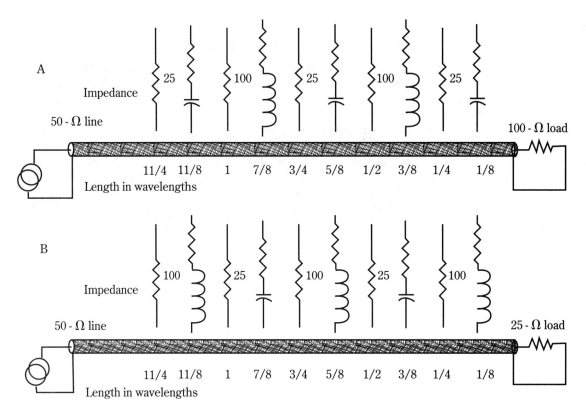

8-2 With a mismatched load, no line length can match the 50-Ω source. With reactive loads, the indicated impedances would, likewise, never be 50-Ω resistive. A. A 50-Ω resistive source, a 50-Ω line, and a 100-Ω resistive load. B. A 50-Ω resistive source, a 50-Ω line, and a 25-Ω resistive load.

mismatch between the line and the load governs the VSWR. Such a measurement will show that line length, line attenuation, and source impedance exert a negligible effect on the VSWR.

These statements and similar discourse on the length of feeder lines, although generally understood, are not mathematically sound. This is because the behavior of transmission lines is more clearly defined in terms of wavelength than by linear length. The reason for this is that the linear length and the electrical length of a line are not necessarily the same. The actual physical or linear length of a line is easy enough to measure, but the electrical length is greatly influenced by dielectric and ferromagnetic substances used in the fabrication of the line. Both exert the same effect: they make the physical length of the line shorter than its electrical wavelength. Thus, a 75-meters line could electrically be 100 meters long. One of the first steps in designing a transmission-line system involves converting physical length to electrical length, which can be expressed in wavelengths or in electrical degrees. In order to do this, the velocity factor of the line must be known.

The velocity factor of transmission lines.

The *Velocity factor* is the percentage of the speed of light that RF energy travels in a transmission line. In lines composed predominantly of air as the dielectric substance, the velocity factor in essentially 1.0, which implies that the physical length and the electrical length of the line are, for practical purposes, the same. Notice that the dielectric constant of air is very close to 1.0, the same as for a vacuum or free space. In such an air-dielectric line, electromagnetic waves travel very nearly as fast as they do in free space.

More generally, the propagation of RF energy along a transmission line is $1/\sqrt{k}$ times free-space velocity, where k is the dielectric constant of the insulating material. Because the dielectric constant of most insulating materials is greater than 1.0, it follows that electromagnetic wave propagation in lines fabricated with insulation is slower than in air or in free space. In turn, this means that the physical length of such a line is less than it would be if calculated on the basis of free-space velocity of the RF energy. Thus, a line with a velocity factor of 0.8 only needs to be 80% of the length that is required if no dielectric material was used in its construction. For example, a line such as described could be 80 meters in physical length, but would then be 100 meters long electrically.

The same reasoning applies to the permeability of any ferromagnetic substance that is used between the conductors of the line. Here, again, any permeability in excess of that of free space (i.e., 1.0) would shorten the line by the factor $1/\sqrt{\mu}$, where μ represents the permeability of the material. Again, the physical length of the line would be less than its free-space wavelength.

It is interesting to ponder that both velocity of propagation and wavelength within a transmission line can be altered by substances affecting the electric or the magnetic components of the RF wave. Yet, frequency remains inviolate—neither dielectric constant, permeability, characteristic impedance, nor dissipative losses affect the frequency. Once set by the source, transmission line parameters cannot change it.

Table 8-4 lists the velocity factors for a group of commonly encountered transmission lines. Multiply the electrical length of a line (in meters) by velocity factor V to get the linear or physical length of the line (in meters). Or, divide the physical

Table 8-4. The velocity factor of various transmission lines. The factor, V, defines the physical length of a line as a fraction or as a percentage of its electrical length.

Line type	Z_0, ohms	Velocity factor, V
open wire, plastic spacers	70–800	0.95–0.97
RG-8/U, SP	52	0.66
RG-8/U, FP	52	0.80
RG-58/U, SP	52	0.66
RG-58/U, FP	52	0.80
RG-59/U, SP	73	0.66
TV twinlead, SP	300	0.82

SP = Solid polyethylene
FP = Foamed polyethylene

length by V to find the electrical length. Notice also that I/V^2 can be substituted for the dielectric constant, k, in the formulas of Table 8-3.

Indicating the effect of reactance in RF systems

The treatment of RF transmission lines, alone, can be a mathematician's delight. On the other hand, many successful practitioners in RF work attain good results via hands-on experience and "flying by the seat of their pants." At the same time, it is also true that much wasted effort and malfunctional operation results because of minimal mathematical understanding. To begin with, it is almost mandatory to acquire some background in ac circuit theory in order to deal with RF systems. If nothing else, the concept of *power factor* should be clearly visualized. It underlies everything else involved in delivering optimal RF power from a source through a transmission line to a load or antenna. The concept of power factor enables you to grasp impedance matching for both low- and high-frequency systems. In addition to power factor, other parameters are used in RF systems to indicate the effect of reactance (Table 8-4).

A source of confusion to students studying ac circuit theory is that pure reactance of either variety (inductive or capacitive) does not consume or dissipate power. In a pure inductance, the current lags the voltage by 90°; in a pure capacitance, the current leads the voltage by 90°. The result of such phase displacements is that no net power is consumed by these devices. Even if this taxes the understanding at first, it is easy enough to remember. It is confusing because the presence of such reactances in the ac circuit causes higher overall power dissipation than would be the case if these reactances were absent. That, however, is because the phase-displaced currents, or the phase-displaced voltages that these reactances produce cause very ordinary I^2R or E^2/R power dissipation in the resistive portions of the circuit. Figure 8-3 shows how the power factor indicates the effect of reactance in a simple 60-Hz circuit.

Another way of looking at this very important circuit phenomena is that reactive elements do, indeed, draw power from the line for one-half cycle. This does not negate previous statements because the consumed power is returned to the line during the next half cycle. So, over a full cycle of operation, neither ideal inductance nor ideal capacitance take any net power. During their half-cycle excursion, however, the currents that they take and return must flow through the resistances of the circuit. In so doing, higher circuit current and higher power dissipation exist in the circuit than would be the case with no reactances present. It could serve a useful purpose to neutralize the effects of unwanted reactance if, as is usually the case, the reactance cannot be physically removed from the system. In RF systems, other parameters than power factor are better suited to indicate the effect of reactance (Table 8-5). The relationship between two such widely used parameters, reflection coefficient and VSWR, are shown in Fig. 8-4.

"Lost" power in a simple feedline system

A list of correspondences, such as shown in Table 8-6, is often encountered in articles that deal with VSWR. A word of explanation in order to prevent some of the commonly held confusions. Notice the list of so-called "power loss" for various

"Lost" power in a simple feedline system 253

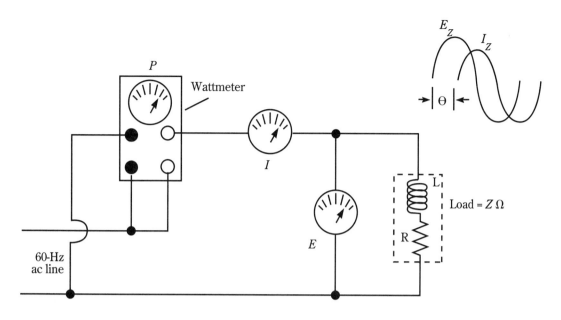

$$\text{Power factor} = PF = \frac{\text{True power}}{\text{Apparent power}} = \frac{P}{E \times I} = \cos \Theta = 1.0 \text{ for pure resistive load}$$
$$= 0 \text{ for pure reactive load}$$

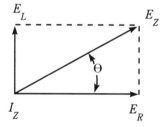

8-3 The power factor conveniently represents the effect of load reactance at 60 Hz. Notice that the "transmission-line" (the connecting wires) do not need to be reckoned at 60 Hz. Also, phase-angle Θ can be readily determined with an oscilloscope.

VSWR levels. The numbers are very approximate, but that is not the matter needing clarification. Just what situation does this table pertain to, and where does the lost power go?

The situation ordinarily pertained to is a simple setup in which the RF source connects to a VSWR indicator, which, in turn, delivers the RF power to the sending-end of the antenna feedline (Fig. 8-5). To avoid complications, assume that the VSWR indicator always reads correctly, and that the antenna feedline is lossless. Under these conditions, the basic cause of any VSWR reading greater than 1.0 is an impedance mismatch between the feedline and the antenna load. The table reveals that

254 Standing-wave ratio and related controversies

Table 8-5. The parameters for indicating the effect of reactance in ac and RF systems.

Parameter	Useful relationships	Comments
Power factor, PF	$PF = \dfrac{P}{E \times I} = \cos\theta$ where: P is true power as indicated by a wattmeter E is load voltage I is load current θ is load the phase angle	Basic relationship that indicates the effect of load reactance. Easy instrumentation at low frequencies. $PF = 0$ for pure reactance $PF = 1.0$ for resistance $\theta = 0°$ for resistance, $90°$ for pure reactance
Reflection coefficient, ϱ (magnitude, only)	$\varrho = \dfrac{V^-}{V^+} = \dfrac{VSWR - 1}{VSWR + 1}$ where: ϱ is Greek letter, rho V^- is reflected voltage V^+ is forward voltage	This basic relationship indicates the effect of load reactance in RF systems. Instrumentation is facilitated by the use of a directional coupler to separate V^- and V^+ $\varrho = 1.0$ for total mismatch $\varrho = 0$ for matched load
Reflection coefficient, $\bar{\varrho}$ ($R \pm jX$ notation or magnitude and phase angle between forward and reflected voltages)	also: $\varrho = \dfrac{Z_L - Z_C}{Z_L + Z_C}$ where: $\bar{\varrho}$ is reflection coefficient expressed as $R \pm jX$ or as magnitude and phase angle. The same is true for load impedance Z_L. Z_C is usually assumed to be resistive.	$\bar{\varrho}$ is particularly useful in solving RF problems because of its direct application to the Smith Chart.
Voltage standing-wave ratio, $VSWR$	$VSWR = \dfrac{V^+ + V^-}{V^+ - V^-} = \dfrac{1 + \varrho}{1 - \varrho}$ where: V^+ is forward voltage V^- is reflected voltage also: $VSWR = \dfrac{1 + \sqrt{\dfrac{P^-}{P^+}}}{1 - \sqrt{\dfrac{P^-}{P^+}}}$ where: P^- is reflected power P^+ is forward power	Reflectometer instruments can provide VSWR readout, which is commonly designated as SWR. Current standing-wave ratio ($CSWR$) is the same as VSWR *Power standing-wave ratio* $= (VSWR)^2$ Reflectometer instruments also provide reflected and forward power readouts. This enables determination of VSWR by calculation or from graph.
Return loss in dB	Return Loss in dB $= 20 \log_{10}\left(\dfrac{VSWR + 1}{VSWR - 1}\right)$ $= 20 \log_{10}\left(\dfrac{1}{\varrho}\right)$	Return loss is largely used in the telephone industry and in formal engineering tests. Return loss is 0 for total mismatch (when VSWR is infinite). Return loss is infinite for a matched system (when VSWR is 1.0).

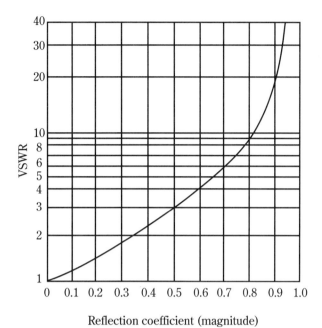

8-4 The relationship between the reflection coefficient and VSWR. Both parameters express the degree of load mismatch.

more power is lost as the VSWR becomes higher. Because the feedline is assumed to be lossless, it is only natural to ponder where the lost power dissipates itself. It, contrary to unfortunate assertions in the technical literature, does not heat up the output amplifier or its tank circuit. It might, indeed, appear to do so, for a destroyed output transistor is often the result of a high VSWR. This will be explained shortly.

Table 8-6. The "lost power" for various mismatches between an RF-source and the line. Although not dissipated, this power is not available for transfer to the feedline.

VSWR	Power "loss" in percent = $\left[1 - \dfrac{4 \times VSWR}{(VSWR+1)^2}\right] \times 100$
1.0 : 1	0
1.3 : 1	2
1.5 : 1	3
1.7 : 1	6
2.0 : 1	11
3.0 : 1	25
4.0 : 1	38
5.0 : 1	48
10.0 : 1	70

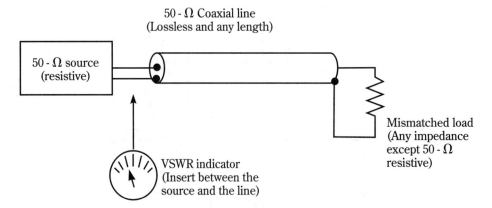

8-5 An RF system that pertains to the "losses" that are listed in Table 8-6.

The erroneous notion that power reflected from a mismatched load enters back into the transmitter would be dispelled if, instead of "power loss," the righthand column of Table 8-6 was labeled "power unavailable for load." This situation is a coupling problem that is caused by an impedance mismatch between the output of the RF source or transmitter and the feedline. Because the transmitter is unable to deliver maximum output into the mismatched impedance it feeds, more power remains within the output circuit of the amplifier. This is what causes the damage; many transistor output stages are simply not designed to tolerate no-load or lightly-loaded operation. Depending on the nature of the impedance mismatch, it could be either an underloaded or an overloaded condition. Both of these conditions can be damaging to the solid-state amplifier.

Reiterating, VSWR monitored at the transmitter infers an impedance mismatch between the transmitter and the feedline. This, in turn, shows that you do not have an optimum coupling situation for delivering maximum available output power to the antenna system. The power often referred to as "lost" is more accurately unavailable via the impedance mismatch at the transmitter as a result of the load mismatch at the far end of the feedline.

Optimum power transfer over a band of frequencies

Consider the simple arrangement in Fig. 8-6. A 2:1 VSWR exists on the assumed lossless feedline and there is no loss of power in this link of power transfer. However, the RF source, being fixed tuned, experiences an impedance mismatch. It, therefore cannot deliver its maximum available power. For the relatively mild VSWR of 2:1, this might not be construed serious in many instances. However, for optimum overall efficiency of the system, the 50-Ω source needs a conjugate match to the feeder line. In the arrangement shown, the length of the feedline can be empirically adjusted for best results, but the perfect source-to-line match cannot be achieved.

Next, imagine that the RF source, instead of being a laboratory RF generator is

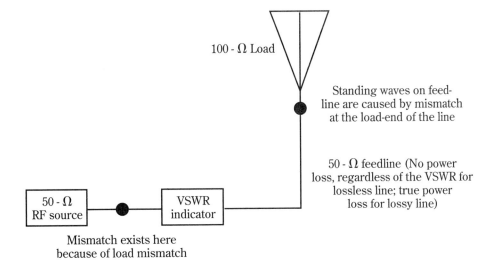

8-6 "Lost" power in a simple RF system. Power is not "lost" at the junction of source and line; it is more accurately, *unavailable*.

a transmitter with a Pi or Pi-L output "tank." This being the case, it is now probable that, for an appropriate length of the feeder line, a true conjugate match can be established by adjusting the transmitter "tune" and "load" controls. The reactance at the sending end of the feedline is simply absorbed in the output resonant circuit of the transmitter. This is fine for low-feedline VSWRs, but it generally will not work when the VSWR is high.

What is needed for high VSWR on the feedline is a transmatch, such as is shown in Fig. 8-7. Although the VSWR remains 2:1 in this illustration, the transmatch is still desirable. First, the transmatch usually dispenses with the need to find a critical length of the feedline. Secondly, this arrangement conveniently enables operation over a band of frequencies, instead of just at the self-resonant frequency of the antenna. With appropriate adjustment of the transmatch, the antenna becomes part of a resonant system that composes the transmatch, the feedline, and the antenna, itself. This enables the antenna to absorb power, despite being driven at frequencies that are considerably removed from its self-resonant frequency. The nice part of this operational mode is that the antenna's radiating efficiency remains high. Third, if the feedline is of good quality (not too long) and the frequency is not too high, the VSWR on the feedline is not of practical consequence. These requirements are not generally difficult to meet for the amateur-radio HF spectrum (from 1.8 to 30 MHz), over which 100-foot lines might be used.

The destination of reflected power

The idea of having high VSWR on an ideally lossless feedline is not easy to accept if you entertain the all-too-common bias against such operation. Even though ideal situations do not exist in real life, moderately high VSWR on low-loss lines yields acceptable performance in everyday practice. The negligible-to-modest losses incurred

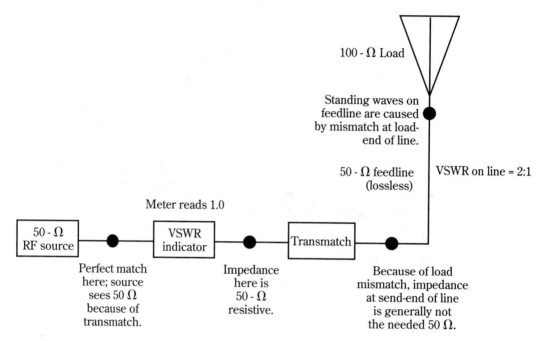

8-7 The use of a transmatch to provide optimum output coupling for the RF source. Transmatch impedance-matching network "fools" RF source into seeing 50-Ω resistance, regardless of VSWR on line.

are certainly a worthy trade off for the advantages of freedom from critical line length, ease of impedance matching the output of the transmitter, and the luxury of operating the antenna off of resonance. Thus, the practically ideal system uses a transmatch at the sending end of the feeder line. An even better arrangement would place the transmatch between the feedline and the antenna, in which case the feeder line would be flat. Not only can a flat line be of any length (consistent with allowable losses), but for any given length, its losses are less than a resonant line. The trouble with this better arrangement is that it is not practical in most installations. Convenience is, indeed, served by having the transmatch at the operator's position.

With reflected-power characterizing resonant lines, you must dispense with the notion that it re-enters the transmitter. Rather, it gets re-reflected and again part of it gets absorbed by the antenna; the remainder again journeys back down the line and again is 100% reflected at the transmitter output. This cycle of events is repetitive until most of the cycling power is eventually absorbed by the antenna. For a lossless line, all such forward and reflected power would find its way to the load. In this process, the interference between forward and reflected energy causes the phenomenon of standing waves. This process also explains why a directional RF wattmeter will indicate that forward power exceeds the actual power output of the source. It is because forward power is assisted by the re-reflected power. The net power transfer, however, is the difference between forward and reflected power.

An illustration of this phenomenon is a ghost image on a TV screen when there is an impedance mismatch between the feedline and the TV receiver. Of course, in this case, the TV antenna is the RF source, and the receiver is the load (Fig. 8-8). Notice the RF energy cycling up and down the feedline. Despite the reflection on the line, the load absorbs most of the energy from the antenna, although not in a desirable way for TV viewing. A different source of ghost images stems from echo-reflection from hills, buildings, or other objects; in such a case, impedance matching in the antenna system cannot remedy the situation.

Multiple reflections due to mismatch between feeder and TV.

8-8 The mismatch between the feedline and receiver produces ghosts. The extraneous images are visual evidence of re-reflections in the antenna system. In this case, the antenna is the RF source and the TV set is the load.

Indirect VSWR estimation at the feedline load end

As already pointed out, it usually is not practical to measure VSWR at the load end of a feedline, particularly if the load is an antenna. Measuring VSWR at the source end of the line would be exactly equivalent if the line was lossless. Real-world feedlines do exhibit losses, however. Although such line losses might not be too large, just from the standpoint of putting a bit less power into the ether, the behavior of the line might be significantly altered. A lossy line will provide inaccurate VSWR readings at the sending (source) end, where the VSWR indicator is inserted. Specifically, the farther away from the load end of the line, the better the VSWR reading will be. For example, if the true VSWR (the load-end VSWR) is 4.5 (for example), you could conceivably obtain a VSWR indication of 2.0 at the source end of the line. Thus, it shows that low VSWR, per se, does not necessarily mean that optimum operation or high radiation efficiency prevails. Instead, the low VSWR monitored at the sending end or source end of the line can be caused by high system losses.

Fortunately, a quantitative relationship exists between line loss and true VSWR. True (load end) VSWR is important because the ability of the lime to withstand high

RF voltage and current is adversely affected by high VSWR. Also, power loss in the practical line increases with VSWR. Finally, the true VSWR enables certain judgments to be made about the operation of the antenna. If, for example, true VSWR is in the vicinity of 15 or 20 instead of 3:6, you might infer that the self-resonant frequency of the antenna is not optimally situated within the desired frequency range or that frequency-range expectations are quite high. A short and/or low-loss feedline would be necessary here.

One way to determine true VSWR from source-end measurements is to first use a cable-loss graph (Fig. 8-9). These curves show the per-100-feet loss for various ca-

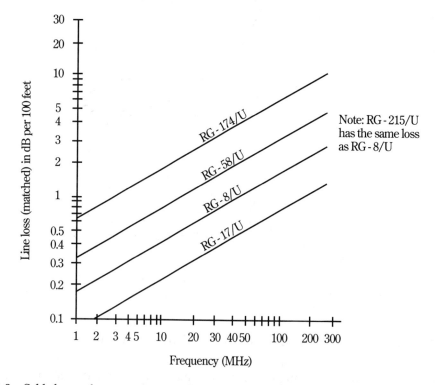

8-9 Cable-loss vs frequency for several coaxial cables. Notice that "cable loss" is conveniently stated for matched-impedance operation on a dB per 100-foot basis.

bles as a function of frequency. Once the loss has thus been established for the particular installation, the true VSWR can be determined by means of the relationships that are graphically depicted in Fig. 8-10. For example, if the matched-line loss has been estimated to be 1.5 dB, and the source-end VSWR has been measured to be 2.5 dB, the load-end VSWR is about 4.3. In using the graph of Fig. 8-9, the matched-line loss per 100 feet is commonly provided by the coaxial-cable manufacturer. Also remember that "VSWR at RF source" implies monitoring VSWR at the sending end of the transmission feedline coming from the load or antenna, not at the junction of the source and a transmatch. However, if a transmatch is used next to the RF source, the VSWR would have to be monitored at the load side of the transmatch. Best results

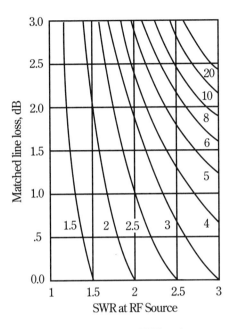

8-10 The curves show discrepancies between VSWR at the source and at the load. The numbers beside the curves indicate VSWR at the load.

would be forthcoming by temporarily dispensing with the transmatch and inserting the VSWR indicator between the RF source and the feedline.

A simple technique to determine line loss

The previously described method to determine the loss in a transmission feedline applies to new lines in good condition. However, after a line has aged and/or has been exposed to the elements, its losses can increase appreciably. This is because of insulation deterioration and sometimes of conducting surface corrosion—especially at coastal locations. A periodic determination of line loss is a worthwhile maintenance procedure. You can, of course, compare RF wattmeter measurements at the source and the load end of the line; the difference between the two is the power lost in the line. Judgment can then be exercised to determine whether this line-dissipated power is an acceptably small percentage of the power delivered from the transmitter.

A more convenient way to ascertain the condition of the transmission line is: place a good RF short at the antenna end of the line. This can be done in many instances without severing the antenna connection. Insert a VSWR indicator between the sending end of the line and the transmitter (do not use a transmatch in this measurement procedure). With the transmitter operated at low power, record the input VSWR to the line. The attenuation of the line (in dB) is then:

$$\text{Line Attenuation} = 10 \, \text{Log}_{10} \left(\frac{\text{VSWR} + 1}{\text{VSWR} - 1} \right)$$

The line attenuation thus derived can be compared to manufacturer's data, taking into account the actual length involved (manufacturer's data is so many dB per 100 feet). In principle, this technique should also work with the load end of the line open-circuited. Capacitive effects of the exposed inner conductor can easily reduce the accuracy of such an open-circuit termination. By the same token, inductance of the shorting element must be minimal when pursuing the technique first mentioned. In any event, the basic idea is that either a perfect short or a perfect open at the load termination produces the known VSWR of infinity. This value enables the line-attenuation equation to work.

The effects of aging and environmental factors tend to degrade a coaxial line's quality more rapidly at higher frequencies. It is, therefore, sometimes prudent to determine and regard line attenuation at frequencies that are several times higher than the actual operational frequency. In this way, you can be warned somewhat ahead of time that you should substitute a new line before trouble from excessive attenuation occurs at the operating frequency.

Direct measurement of VSWR

The voltage standing-wave relationships in Table 8-7 are based on the entities responsible for actually causing standing waves to exist on a transmission line. These entities are the forward (incident) voltage, $V+$, and the reverse (reflected) voltage, $V-$. It is, fortunately practical to monitor or measure these two voltages because of the selective action of the directional couplers used in a VSWR indicator. There is, however, another way to determine VSWR that is particularly useful at UHF and at microwave frequencies.

The interaction between the incident and the reflected voltage waves produces the standing-wave phenomenon, which can be directly measured with an RF voltmeter. In so doing, you need not be mindful of the causative agents, $V+$ and $V-$. The RF voltmeter simply responds to the interference product (i.e., to the actual voltage standing wave). Because no directive couplers are used with the RF voltmeter, there are no separate responses for $V+$ to $V-$. Let the voltage thereby monitored be designated by E to distinguish it from its two constituent voltages, $V+$ and $V-$. If the RF voltmeter is moved along the line, the detected voltages will vary from a maximum to a minimum value, that is, from E_{max} to E_{min}. Each of these two values will be repetitive at ½-λ intervals along the line. This provides a neat way to calculate the VSWR; it is simply E_{max}/E_{min}. Notice that E_{max} and E_{min} are always spaced ¼ λ apart.

Because of physical considerations, the direct method of measuring VSWR is best suited to the UHF and microwave regions of the spectrum. "Slotted line" sections with calibrated sliding probes are specially made for such evaluations. The system transmission line is interrupted by the slotted line, and readings are recorded under actual operating conditions. Although the microwave diode used as detector has a square-law response, this is taken into account in a calibration chart. If the VSWR is thereby measured in dB, conversion to actual VSWR can be made via the relationship:

$$VSWR = \text{antilog}_{10}\left(\frac{dB}{20}\right).$$

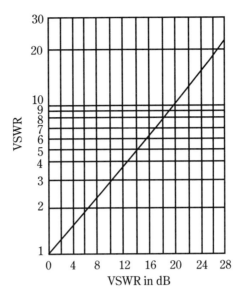

8-11 A conversion chart between VSWR as a ratio and VSWR in dB.

This derives from $dB = 20 \log_{10}$ (VSWR). The chart of Fig. 8-11 is also applicable.

No calibration of the probe meter is needed if a variable attenuator is inserted between the RF source and the slotted line. E_{max} then is made to correspond to a full-scale reading. The dB change removed from the variable attenuator to restore the full-scale reading when the probe is at E_{min} (the VSWR in dB). The arrangement is shown in Fig. 8-12. The conversion formulas and chart alluded to apply here also.

An even simpler VSWR determination can be made from the basic arrangement of Fig. 8-12, if the probe meter is calibrated in volts. In this case, the variable attenuator is not needed. The attenuator, however, lends flexibility to the measurement process, and it is often good to follow the RF source with at least 6 dB of attenuation as a buffering or isolation technique. From the fundamental concept of SWR, the following relationships apply:

$$VSWR = \frac{E_{max}}{E_{min}}$$

where E_{max} and E_{min} are obtained by sliding the probe along the slot. Both readings must be taken with the same probe and probe-meter adjustments.

Also, $VSWR$ in dB $= 20 \log_{10} \left(\dfrac{E_{max}}{E_{min}} \right)$

Both VSWR measurement techniques yield VSWR "from the viewpoint of the RF source," as is commonly done in amateur-radio practice at lower frequencies. If, however, the true VSWR at the line-load junction is desired, it is often feasible to insert the slotted-line section immediately ahead of the load.

Remember when using the slotted line that the probe does not contact the in-

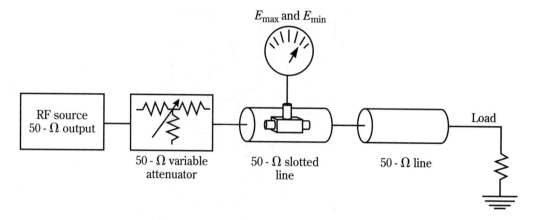

1. Adjust the system for full deflection of the probe meter when it is positioned at E_{max}.

2. Move probe to E_{min}, then adjust attenuator to regain full meter deflection.

3. VSWR in dB is the *difference* in attenuator settings in steps 1 and 2.

4. VSWR in dB can be converted to a numerical ratio. See text.

8-12 A slotted-line arrangement to determine VSWR. Implementation is practical for UHF and microwave frequencies, but the concept is instructional for all RF systems.

ner conductor of the coaxial structure. Rather, it protrudes just sufficiently to sample the electric field. Indeed, the accuracy of measurements is enhanced with the least probe "invasion" needed to make measurements. However, as previously alluded to, E_{max} and E_{min} must be made with the same probe adjustment. The reason for minimizing probe insertion is that the probe not only samples the electric field, but also acts as a mild impedance discontinuity, and it can thereby affect the measured VSWR.

Standing waves do not have a sinusoidal-like shape, as is sometimes depicted. Rather, the E_{min} regions tend to be sharper than the E_{max} regions (Fig. 8-13). Measurement accuracy can often be enhanced by establishing the location of E_{max} as halfway between two adjacent E_{min} locations.

Pitfalls in transferring RF power

The following are the commonly encountered difficulties in transferring RF power from the source to the load. Often, the RF source is a transmitter and the load in an antenna. Enough distance separates source and load so that such factors as line attenuation, line VSWR, and line radiation might have to be dealt with. Other RF sources, delivery methods, and types of loads are also encountered in industry, research, and experimentation.

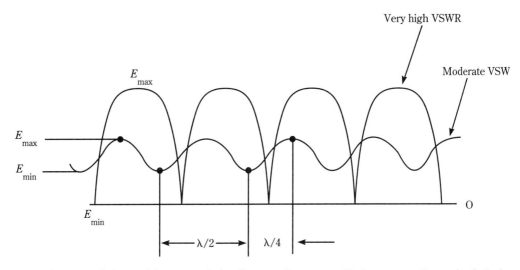

8-13 The general shape of the transmission line standing waves. Notice narrow E_{min} and relatively broad E_{max} regions.

- Transmitter with 50-Ω output is followed by 50-Ω VSWR indicator, and a 50-Ω coaxial line feeds a resonant antenna with radiation resistance of 50-Ω (Fig. 8-14). The VSWR indicator, which should read 1.0 in this arrangement, reads considerably higher. What is the possible cause of such departure from theoretical performance?

  A likely reason that the ideal 1.0 VSWR is elusive in this basic arrangement is that the transmitter output contains considerable harmonic energy.

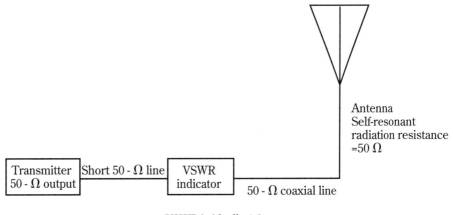

8-14 A higher than 1.0 VSWR is sometimes observed in a "perfect" arrangement. Harmonic energy reflected by the antenna is the common culprit. The usual remedy is a filter to follow the transmitter.

Many antennas will not absorb the harmonic energy, which gives rise to reflected power, and therefore to a higher-than-unity reading on the VSWR indicator. This undesirable mode of operation can be confirmed by substituting a 50-Ω dummy load for the antenna. Although a nonunity VSWR reading might not, in itself, signify serious operating trouble, the harmonic energy can cause serious communications interference and TVI. A low-pass filter at the output of the transmitter is a usual remedy for this situation.

Another remedy is possible for trouble of this kind. The presence of RF on the outer surface of the coaxial line tends to disturb the reading of the VSWR indicator. A balun might be needed at the antenna feed point. Or, the physical take-off of the line, relative to the antenna might allow such RF pickup. Also, the station operating position might be too close to the antenna. Notice, however, that it is not true that a line will exhibit this malfunction just because of a high VSWR—operating an antenna off its frequency of self-resonance, for example. Suitable grounding techniques at the transmitter might alleviate this trouble in some cases.

A transmitter with 50-Ω output operates into a VSWR indicator, followed by a transmatch, the feedline, and the antenna, which is generally driven off of its self-resonant frequency (Fig. 8-15). It is sometimes observed that very little power is radiated from the antenna, even the tune-up process produces the sought-after low VSWR reading. Various checks show that the transmitter, the SWR indicator, the transmatch, the feedline, and the antenna are all without defect. What, indeed, goes on here?

This is an interesting dilemma. With a low-power transmitter, the trouble spot can be elusive. With high power, troubleshooting can be faster if you analyze the reason for the overheated transmatch. What has happened is that the unfortunate adjustment of the transmatch tuning controls enabled it to operate as an absorption wavemeter. In such an operational mode, the RF power is largely dissipated in the resonant circuits of the transmatch, and very little is released to the antenna feedline. Thus, it is wise to record proper transmatch settings for different transmitter frequencies. Although you should recognize telltale signs of transmatch maladjustment after you have had much experience, when the adjustments must be quickly made, it can be easy to make an erroneous judgment. This is especially true because the

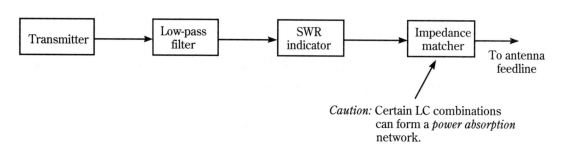

8-15   A functional-block arrangement for conveying RF power to the antenna. The impedance-matcher or (transmatch) can sometimes be maladjusted to behave as a dissipative load.

VSWR indicator might convey the fact that the transmitter is "happy" with the situation by showing a low VSWR.

There is a more subtle reason why optimum adjustments of the transmatch should be logged after these are leisurely made and thoroughly investigated. You can obtain reasonable results for two or several combinations of settings of the tuning controls on the transmatch. One way you can determine which combination is optimum is to insert an RF ammeter in the line following the transmatch. It does not matter that the indicated RF current might vary with the line length; you are interested in relative deflections, not an absolute reading of RF current. With the RF ammeter in place, choose the combination of transmatch tuning adjustments that results in the highest line current. In most instances, this can be expected to be that combination which uses minimum inductance. In any event, if everything is in order, maximum RF line current should coincide with minimum reflected power and/or minimum VSWR at the transmitter output.

Another reward for logging transmatch tuning adjustments and VSWR indications is that it provides a clue to future changes in overall performance. If the adjustments or readings undergo change, you can suspect deterioration of coax insulation, corrosion of connectors, approach of trees or objects to the antenna, effects of ice or water, and so on. In antenna systems where grounding is important, the RF integrity of the ground system should be investigated following any need to change the tune-up controls. In systems that use grounded antennas with no transmatch, the VSWR monitored at the transmitter can actually improve as the ground resistance increases. This can be deceptive because a worsening ground reduces the radiating efficiency of such an antenna.

- An RF source is coupled to a resistive load via a low-loss transmission line that is many wavelengths long. A considerable mismatch exists between the load and the line, which causes a high VSWR. The actual length of the line is adjusted so that a partial impedance match exists between the source and the line. This is indicated by an RF wattmeter between the source and the line (Fig. 8-16). Such an arrangement can be expected to be frequency sensitive because changing the frequency effectively changes the electrical length of the line. This, in turn, varies the sending-

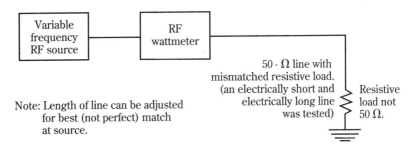

8-16 An arrangement that exhibits frequency sensitivity, despite the resistive load. This is a compromise power-transfer situation at best. The longer the line, the greater the frequency sensitivity.

end impedance of the line, and thereby changes the source-coupling situation. However, the frequency sensitivity of a previously tried similar arrangement with a much shorter transmission line was much less severe. Wherein lies the performance discrepancy between the two arrangements?

This situation is cited in order to illustrate a characteristic of resonant lines (lines with high SWRs) that is not always realized. The situation has been simplified by assuming a purely resistive mismatched load, whereas antennas became reactive on either side of self-resonance. The point is that the longer the line and the greater the VSWR, the greater is the frequency sensitivity of the system. This implies that a very small frequency change will greatly upset the output-coupling situation that is obtained by adjusting the line length at the original frequency. The source-to-line impedance relationship repeats every half wavelength of line, but the tolerance to changes in frequency grows progressively worse with the number of half wavelengths of transmission line. If an electrically long line must be used, the logical way to reduce frequency sensitivity would be to make the load resistance more nearly the value of the characteristic impedance of the line. Indeed, with a matched load resistance, the line would be flat and frequency sensitivity would disappear. Also, the line could then be of any length. In any event, remember that adjusting line length alone doesn't change the line VSWR, and at best, can only provide a partial match to the source.

Considering the comparison of 40 or 80 meters (for example) with the size of an average city lot, the previous example is likely to have more relevance to VHF, UHF, and microwave frequencies than to the HF amateur radio bands. Because the frequency characteristics of an antenna would make the overall system performance even more critical, with respect to frequency, wide-band capability requires either a "trombone" (a segment of line that is adjustable in length) or a transmatch. These devices enable re-establishment of the best power transfer each time that the frequency is changed. For optimum power transfer, however, a transmatch must be used in order to enable a true conjugate match between the source and the line.

- The liberal use of the terms "impedance" and "resistance" can be responsible for difficulties in optimizing the performance of RF systems. "Characteristic impedance" of transmission lines, and the feedpoint "impedance" of an antenna are used frequently. Remember that the transmission-line impedance is resistive anywhere along its length when it is terminated by a certain-valued resistive load. The corollary of this statement becomes one of the practical definitions of transmission-line characteristic "impedance." That is, the load resistance that causes a transmission line to be flat (to have no standing waves) is the characteristic impedance of the line. For most practical purposes, this 50-Ω, 73-Ω, 300-Ω, or other impedance is almost a pure resistance. At the same time, it is not dissipative—it wastes no power. Power is wasted in practical lines by the ohmic resistance of the conductors and by imperfect dielectric and insulating materials, but not by this entity called *characteristic impedance* (more practically, *characteristic resistance*).

Now consider the terminology of the load, usually the antenna. It is commonly stated that the self-resonant impedance of an antenna is so many ohms. What this actually means is that the impedance "seen" at the feedpoint of an antenna current loop is a resistance, that is, the radiation resistance. Lest it appear that I am focusing too much on the fine points of academic trivia, contemplate this rather common trouble:

An antenna is altered until its feedpoint impedance is 50-Ω at a desired frequency. However, when fed by a long 50-Ω coax line, high VSWR exists on the line. Some degree of transmitter loading might be attained by adjusting the line length, but proper operation is not forthcoming by this technique. If much power is involved, an appreciable temperature rise occurs on portions of the line. And, the radiation efficiency of the antenna is not what it should be. What can be wrong with this simple arrangement?

The antenna, of course, is mismatched to the feedline and high dissipative losses occur in the line because of its high VSWR. For the desired match to exist, the line must "see" a 50-Ω resistive load, not just reactance-resistance combinations that happen to be 50-Ω impedances. To bring this about, impedance-matching techniques not only cancel reactance, but affect the appropriate transformation between nonequal line and load resistances. Impedance matching is needed because a 50-Ω impedance load is generally a mismatch for a 50-Ω line. The exception is when the 50-Ω "impedance" has zero reactance and is therefore resistive.

It is important to understand the true significance of this poorly operating RF system because very useful performance is attainable with high VSWR on the transmission line; it is not necessarily the high VSWR that causes inefficient operation, granting that a combination of high VSWR and lossy line must not be allowed to sap too much power. With a lossless line, no power would be lost, regardless of VSWR. Under the right circumstances, virtually all available power from the transmitter can be absorbed and radiated by the antenna—even though a moderately high VSWR exists on the line, and the line exhibits some, but not excessive, losses.

The right circumstances did not exist in this example. What would have been necessary is an antenna-radiation resistance in the vicinity of 50-Ω. This, of course, could only exist at the frequency of self-resonance. However, such operation is too restrictive in amateur-radio practice—it is ordinarily desirable to attain efficient RF power transfer to the antenna over a band of frequencies. The ideal way to accomplish this is to insert a tunable impedance-matching network (transmatch) between the feedline and the feedpoint of the antenna. With proper adjustment of the transmatch, the feedline would be flat (no standing waves) and the sending end of the feedline would be 50-Ω resistive, regardless of its length.

Unfortunately, though ideal, such an arrangement is generally not practical. In practice, the transmatch is inserted between the transmitter and the feedline. In such an arrangement, the transmitter is still faced with the needed resistive 50 Ω for efficient power transfer. The antenna system (the transmatch, the feeder line, and the antenna) is resonant. Thus, the antenna, which, of its own accord would not accept much power at frequencies other than its self-resonant frequency, now accepts full power over a range of frequencies. It does this because the transmatch cancels the effect of antenna reactance. The antenna reactance is primarily responsible for its "natural" refusal to operate over a band of frequencies. Frequency changes require transmatch retuning.

Generally, the radiation resistance of an antenna changes less with frequency change than does its reactance. This makes wideband operation easier, but in any event, re-adjustment of the transmatch can compensate for both reactance and resistance variations in the antenna. Notice that the transmatch does not "tune" the antenna. Rather, it provides a conjugate impedance match between the transmitter and the feeder-antenna combination.

- An impedance mismatch exists at the junction of the feedline and the antenna. It is proposed to experimentally find the length of 50-Ω line that will present the desired 50-Ω resistance at its sending end so that the transmitter will deliver maximum available power. What is wrong with this strategy?

    This technique has, unfortunately, received support in the technical literature; it has retained a measure of credence over the years because sometimes a degree of success is attained. However, closer scrutiny of performance reveals that by varying feedline length, you can get a sending-end impedance of 50 Ω, plus some reactance. For a different length, you can get a pure resistance value other than 50 Ω. Tantalizingly, no feedline length will yield the sought-after 50-Ω resistive input. Assuming that no transmatch is used, this matching problem has two practical remedies. First, the line of appropriate characteristic impedances, Z_c (¼ λ long electrically), can be designed to present the needed 50-Ω resistance to the transmitter when working from a non-50-Ω radiation resistance (Fig. 8-17). The changes of accidentally encountering this situation are slim. $Z_c = \sqrt{Z_a \times Z_b}$, where Z_a and Z_b are the resistances transformed by the ¼-λ matching line. Second, a critical length of line can sometimes be resented with the needed conjugate-match by the tank circuit of the amplifier itself. This matching technique is sufficient only for low-line VSWRs.

- It is found that different VSWR readings can be obtained along the length of a mismatched transmission line. Is this an acceptable situation?

    The answer here is yes and no. On a lossy line, it will always be true that the VSWR will become progressively lower as it is monitored farther away from the mismatched load. With a lossless line, the VSWR is the same anywhere along its length. Assuming a practical line with negligible loss, this will remain very nearly so. In practice, you might find that considerable variations in VSWR are indicated. This might be an instrumentation error as a result of impedance variations along the line. Regardless of the characteristic impedance of the line, a VSWR of 4 will produce an impedance variation of 16 every ¼ λ. A VSWR of 5 will result in impedance variations of 25. Whereas VSWR indicators might produce reasonably accurate readings for VSWRs under

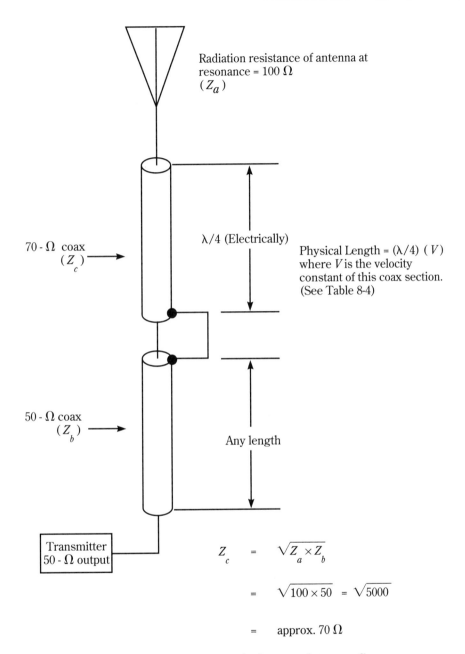

8-17 An example of a single-frequency matched system that uses a line.

three, they might not be trustworthy at higher values. In practice, if a considerable change in VSWR is indicated over a short length of line, the higher reading tends to be the most accurate. "Short length of line" should be interpreted as plus or minus a ⅛ λ or less.

- RF induction-heating applications, because of the widely varying loads, tend to be abusive to the RF source. For the sake of operating convenience, optimum tuning adjustments are often compromised. In the arrangement in Fig. 8-18, satisfactory heating of a work piece is achieved under the indicated conditions, which correspond to a VSWR of 3.0. What demand is made on the RF source, with regard to power-output capability?

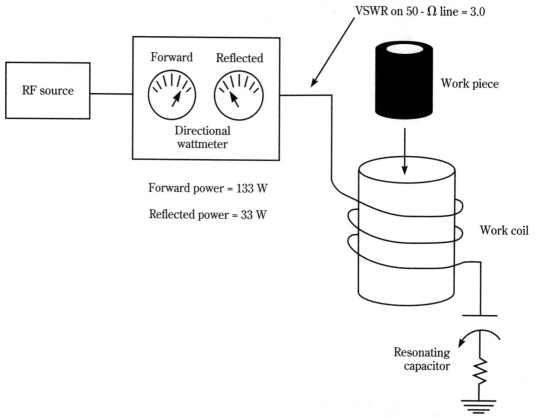

8-18 RF induction heating often operates with mismatched conditions.

The pitfall to avoid here is that the RF source is not called on to provide 133 W to the load system. The net power consumed by the load and by line, work-coil, and capacitor losses is 133 (33 or 100 W). The "extra" 33 W does not come from the RF source, but rather from re-reflected reflected power. This is reminiscent of apparent power in a low-frequency ac circuit, which is always greater than the true power when the power factor of the circuit is not unity.

Having said this, there are other valid reasons for incorporating a sizable safety factor in the design and specification of the RF source that is used for this type of service. You must take into consideration the duty-cycle of the operating procedure, as well as the VSWR if the RF source is solid-state. Thus, a source with a power output capability of 150 W or more might be used in the practical implementation of the system. Nonetheless, the depicted operating conditions only deliver 100 W into the load system. What happens is that the applied power is split into the two components, forward power and reflected power, and the unique selective feature of the directional wattmeter enables these two components of power to be separately displayed.

Contrary to statements in technical literature, the reflected power is not fictional. Its real existence is not only confirmed by the previously described behavior, but it is also evidenced by its ability to dissipate itself as heat in the line and system losses. It does not, however, re-enter the RF source, and it is not correct to attribute damage to the RF source by this allegation. The damage is caused by the change in loading impedance (coupling) that accompanies VSWR and reflected power.

- A 300-MHz RF system is carefully breadboarded and measurements of performance are made with good-quality instruments. The load VSWR is about 1.8, which is deemed satisfactory. This performance was not duplicated during the first production run; it was found that considerable revamping of the layout was necessary to again achieve a maximum load VSWR of 1.8. What might have gone wrong?

Although it is true that inordinate skill might be needed to translate breadboard performance to the finished product at VHF and UHF frequencies, an often-overlooked contributor the described dilemma is to ignore the effect of the inserted wattmeter or VSWR indicator. At lower RF frequencies, this several inches of additional line has a negligible effect. Even at UHF frequencies, there is no effect if the VSWR is 1.0, which corresponds to the practically elusive perfect match. At 300-MHz and with the acceptable mismatch manifested by the 1.8 VSWR, performance can be appreciably different with the meter in and out of the system.

Fortunately, manufacturers of high-quality RF instruments provide data to correct for, or better still, to circumvent the line-lengthening effect of making such high-frequency measurements. The scheme amounts to adding a prescribed length of 50-Ω coaxial line to the meter so that the total line-lengthening that results from the measurement process is ½ λ. Under such measurement conditions, virtually no disturbance is imparted to the RF system that is being measured. This is because of the duplicating and repetitive nature of ½-λ sections of transmission line. Of course, the velocity factor of the added line must be taken into account. What counts is the electrical length, not the mere physical length of the added section.

To simplify matters, manufacturers provide charts, such as for the popular Bird Model 43 Thruline directional wattmeter (Fig. 8-19). Because the meter looks like four or five inches of 50-Ω transmission line, its effect on measurement accuracy below 100 MHz or so, is negligible. However, at the 300-MHz operating frequency of this system, the chart shows that 10 inches of RG-8/U coaxial cable must be added to the meter. This is because the five inches of equivalent line length for the meter is not a negligible part of a ½ λ at 300 MHz. The use of the chart (Fig. 8-19) is facilitated because the manufacturer has already taken into account the velocity factor of the added cable; simply add the indicated physical length. Be sure, however, to include connectors in the length measurement.

- Why are VSWR readings less useful to the antenna designer and experimenter than to the station operator?

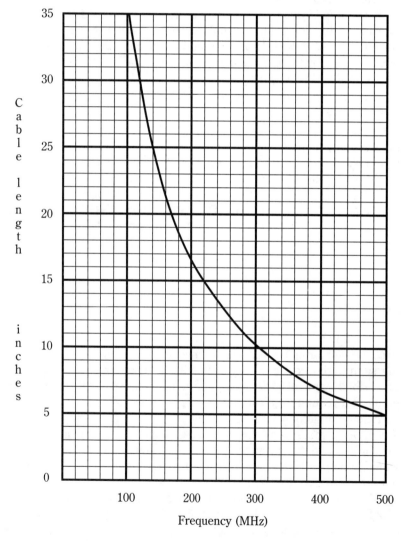

8-19 The required RG-8/U cable length for the Bird 43 to equal ½ λ. The Bird 43 directional wattmeter has long been recognized as a reference instrument. The line-length correction technique does not need to be used below about 100 MHz.

By observing and logging VSWR data, the station operator can glean circumstantial evidence of change in the antenna system. Such things as moisture, dimensional changes, ice, approach of objects (such as trees), deteriorating insulation, opens or shorts in feeder lines or in resonant traps, variation in ground conductivity, balun-core saturation or physical breakage, connector malfunctions, etc. are accompanied by changes in the VSWR reading. Indeed, even an improved VSWR reading from the previous reading should be investigated; increased ground or feedline losses can, unfortunately, deceive the unwary operator by causing lower VSWR indications.

For the designer or experimenter, VSWR data can be inadequate or ambiguous. Consider such information as a VSWR of 4.0. Assuming use of a 50-Ω feedline to the antenna, such a VSWR does not show whether the radiation resistance of a resonant antenna is 12.5 Ω or 200 Ω. Even worse, if the antenna is operated off of resonance, the 4.0 VSWR does not show whether the antenna is inductive or capacitive; if you want to lower the VSWR, you do not know whether the antenna is too long (inductive) or too short (capacitive).

Accordingly, it is much more useful for the antenna designer and experimenter to be able to work with impedance data in the form of resistance plus reactances: $R \pm jX$. Such data is attainable directly at the feedpoint of an antenna via such instruments as a vector impedance meter; a network analyzer, a reflection-coefficient bridge, or an R-X noise bridge. The R-X noise bridge is particularly useful because it is simple, inexpensive, small, and readily battery operated. When you have $R \pm jX$ data over a span of frequencies, it then is easy to find what type of modifications are needed to achieve desired results. Not only does it become evident whether the antenna should be shortened or lengthened (a resonant antenna displays zero reactance), but its radiation resistance can be fine tuned by its height above ground. By the same token, the transformation ratio of a balun is clearly indicated. Finally, the $R \pm jX$ information can be readily used to calculate unambiguous VSWR information.

An example is how VSWR can be calculated from resistance and reactance data obtained at the center feedpoint of a dipole antenna by using an R-X noise bridge. For example, this measurement yields a resistance of 30 Ω and an inductive reactance of 40 Ω; the measurement is being made at 4.0 MHz. It is best to calculate the VSWR at the junction of the antenna feedpoint and to use 50-Ω coaxial cable as the feedline.

Because the reactance is inductive, the resonant frequency of the antenna is lower than 4.0 MHz. Aside from this interpretation, it does not matter whether the reactance is inductive or capacitive insofar as the VSWR to be calculated is concerned.

Although the formula for VSWR is simple and straightforward, be careful with the arithmetical and algebraic operations.

It is interesting to contemplate that if you had only known that the VSWR at the antenna-coax junction was 3.00, there is no way that the problem could

$$VSWR = \frac{(Za + Zc) + (Za - Zc)}{(Za + Zc) - (Za - Zc)}$$

where: Za is $30 + j40$ in this example.
Zc is $50 + j0$, the characteristic impedance of the coaxial cable.

$$VSWR = \frac{(30 + j40 + 50 + j0) + (30 + j40 - 50 + j0)}{(30 + j40 + 50 + j0) - (30 + j40 - 50 + j0)}$$

$$= \frac{(80 + j40) + (-20 - j40)}{(80 + j40) - (-20 - j40)} = \frac{\sqrt{80^2 + 40^2} - \sqrt{-20^2 - 40^2}}{\sqrt{80^2 + 40^2} - \sqrt{-20^2 - 40^2}}$$

$$= \frac{\sqrt{8000} + \sqrt{2000}}{\sqrt{8000} - \sqrt{2000}} = \frac{89.44 + 44.72}{89.44 - 44.72} = \frac{134.16}{44.72} = 3.00$$

have been "worked backwards" to yield the resistive and reactive parameters of the antenna.

- Because of various practical considerations, a designer of an RF system is ready to choose between a load with zero reactance, but with twice optimum resistance and a load with twice the reactance of the characteristic impedance value of the transmission line, but with optimum resistance. The designer is about to flip a coin. What pitfall awaits?

In both instances "optimum resistance" is the value that corresponds to the characteristic impedance of the transmission line. Assuming that the line can be considered lossless (because of favorable conditions that pertain to its length, quality, and operating frequency) the designer's "common-sense" approach is that one of these loads is about as good as the other. Unfortunately, this answer is far from true. It turns out that tolerance to a departure from optimum load resistance is much better than tolerance of the same number of ohms of load reactance, which absorbs no power. The affected parameter is the all-important available power that the source is willing to deliver into a mismatched load.

The curves of Fig. 8-20AB depict the operating conditions for the previously mentioned mismatches. From curve A, you can see that the load with twice its optimum resistance still allows nearly 90% of the maximum available power from the source to be delivered to the load. In contrast, curve B reveals that the load that has a similar number of reactive ohms can only receive 50% of the maximum available power from the source. This is the concept of "lost power" that was previously discussed.

The curves of Fig. 8-20 AB pertain equally for any given ratio and its reciprocal. Thus, a ratio of 2 of load resistance to characteristic impedance of the line is the same as a ratio of ½. The same logic applies to 3 and ⅓, 4 and ¼, etc. For ease of reading this graphical data, it is always best to convert the fractional ratios to whole-number ratios.

A secondary reason why the zero-reactance situation is favored by the source often enters the picture when operating antennas off their self-reso-

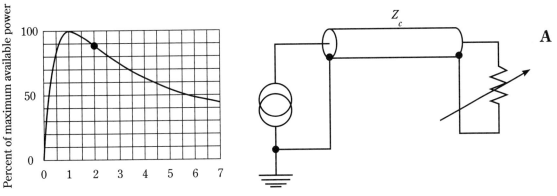

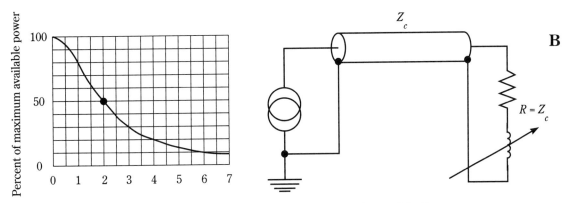

8-20 A comparison of load resistance and reactance on available power from the source. A. A variable load resistance with zero load reactance. B. Variable load reactance with load resistance equal to the characteristic impedance of the line. The curves show that the power-limiting effect is much greater for reactance than for resistance.

nant frequency. The radiation resistance of the off-resonant antenna usually does not change as rapidly with frequency as does reactance. This tends to simplify impedance matching and to make it less critical.

Although variable inductive reactance is depicted in the representative circuit of Fig. 8-20B, the situation would apply in the same manner to the same number of ohms of capacitive reactance.

Practical measurement of coaxial cable velocity factor

More often than not, the specified velocity factor of coaxial cable will be somewhat different in practice. Because RF elements, such as antennas, transmission feedlines, tuning stubs, and resonators must, in many systems, be predetermined electrical wavelengths, the corresponding physical length can only be as precisely measured as the accuracy with which you know the frequency, the physical length, and the velocity factor. These three parameters are bound to either in a single equation:

$$f = \frac{492 \times VF}{L}$$

where:
f is frequency in MHz
L is length in feet
VF is the velocity factor of the coaxial cable

- If a digital frequency counter is used to measure frequency, excellent precision results for use in this formula. Then, with a bit of care exercised in measuring the physical length of a section of cable being investigated, a very good value of actual velocity factor can be obtained. For this calculation, this formula can be conveniently used in the following form:

$$VF = \frac{f \times L}{492}$$

- The practical procedure for determining velocity factor derives from the simple arrangement in Fig. 8-21. Here, a section of the coaxial cable is selected for testing. About 50 feet is a suitable length, but this length should be measured as carefully as possible and recorded. Notice that the far end is short-circuited. The RF generator is tuned through the 6.5- to 8.5-MHz frequency range in order to observe the occurrence of a null. The null occurs at the frequency that corresponds to a true electrical ½ λ of the line. This is because at the frequency that corresponds to an electrical ½ λ, the input of the line appears exactly as the output, which is a short-circuit. Then, the null frequency and the measured physical length are used in the formula to calculate the velocity factor.

- Nothing is sacred about the suggested 50-foot length; other lengths will produce the null at other frequencies. However, it is important to find the lowest frequency that will cause the null in the test arrangement. To get an approximate idea of this frequency, first make a calculation using an assumed velocity factor of 0.7. The true frequency will be somewhere near this approximate value; it will not be three or five times greater, corresponding to odd harmonics of f.

The R-X noise bridge

Paradoxically, a very useful RF instrument is among the simplest and least expensive of the various measurement devices. The R-X noise bridge is inordinately useful because it, unlike the VSWR indicator so commonly encountered, provides full impedance information in an RF system. For example, you can easily obtain the resistance and reactance of an antenna. Moreover, the type of reactance, whether in-

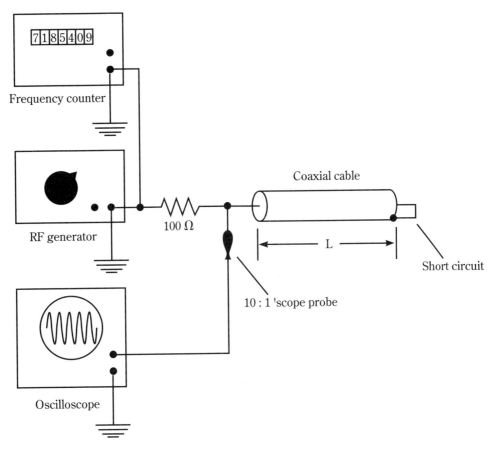

8-21 The arrangement to determine the velocity factor of coaxial cable. When the lowest frequency that corresponds to a null is found, the section of coaxial cable is electrically one-half wavelength. This, then, allows calculation of the velocity factor.

ductive or capacitive, is readily indicated during the measurement process. Armed with such data, you can quickly ascertain whether the antenna is too long or too short. Also, the radiation resistance at resonance is, at the same time, available. Convenient determination of these 15 parameters are very important because assumptions commonly made are, more often than not, considerably different because most antennas do not operate in the ideal free-space environment.

The R-X noise bridge can provide measurements of more than sufficient precision for most practical purposes. The null detector of this bridge instrument is intended to be a modern communications receiver, in which case the frequency of the null can be read on a digital readout instead of from analog dial calibrations. The R-X noise bridge is battery operated. For greatest convenience, the receiver that is used as a null detector should also be capable of battery operation. Most measurements are not time consuming, and excessive demand does not need to be made on either the noise-bridge battery or the receiver battery.

The Palomar Model RX-100 R-X noise bridge is shown in Fig. 8-22. Notice that there are inputs for the unknown impedance and for the receiver that is used as the null detector. The pointer knob that is designated as "R" is straightforward enough; it indicates the resistance component of the unknown impedance when a noise null is attained at the receiver. The other pointer knob actually indicates plus and minus capacitance in one of the bridge arms. Because bridge balance exists at midposition (70 pF) of a 140-pF variable capacitor, departures from balance correspond to a range of 70 pF on either side of the balance. The significance of these capacitance readings is that, knowing the frequency from the receiver, you can easily calculate either inductive or capacitive reactance, or can readily obtain these reactances from a chart.

8-22 The Palomar RX-100 R-X noise bridge. This simple device enables convenient determination of resistance and reactance throughout the 1- to 100-MHz frequency range.

In practice, the frequency of the noise null is generally more important than having a precise reading of the resistive and reactive components of the unknown impedance. That is why the pointer-know calibrations do not suffer for lack of finer resolution.

The functional diagram of the R-X noise bridge is in Fig. 8-23. The Palomar Model RX-100 instrument does not use a square-wave modulator. This optional function block is used by some manufacturers to enrich the electrical noise spectrum. However, it appears that entirely satisfactory results can be realized by appropriately selecting the Zener noise-generating diode, and by sufficient amplification. In particular, low-voltage Zener diodes work well because their breakdown mechanism is predicated more on avalanching than on Zener action.

The generated noise spectra encompasses an extremely wide frequency band. This enables the Palomar Model RX-100 to be useful from 1000 kHz to 100 MHz. At any given frequency, the transient RF supplied by the Zener diode can be

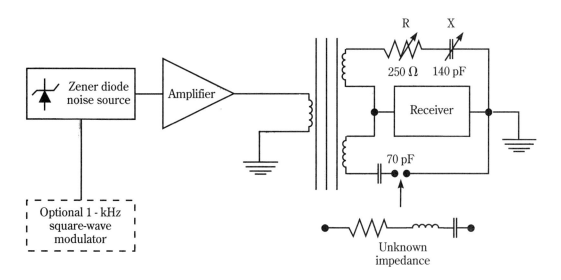

8-23 A functional diagram of the R-X noise bridge. Chopping the noise spectrum with an audio frequency square-wave is claimed to facilitate detection of the null, but it is not a basic necessity.

considered to be both amplitude and frequency modulated, both in random, but rapid, fashion. That is why, even though the receiver responds to a very narrow "slice" of this spectrum, it comes through as electrical noise. When the bridge in balance, cancellation of the noise supplied to the receiver occurs. This constitutes the null, at which time the receiver frequency and the readings of the two pointer knobs should be recorded.

An AM receiver is beat and the noise level can be conveniently monitored from the receiver's S meter. Conversely, the noise might be listened to in the speaker; if this method is used, the AGC should be off. An ac voltmeter connected across the voice coil of the speaker is also a satisfactory method.

In using the device, the connection between the bridge and the receiver can be coaxial cable of almost any length or characteristic impedance. The connection between the bridge and the unknown impedance should be as short as possible. If coaxial cable is used, it should be no longer than a small fraction of a wavelength. An exception is a rather complicated use in which the noise bridge "looks at" the sending end of the antenna feedline. This measurement technique requires the additional use of a Smith Chart and tends to negate the simplicity with which most data can be acquired with the use of the noise bridge and receiver.

Ordinarily, antenna information is obtained by using the noise bridge at the current loop of the antenna, with the feedline temporarily disconnected. This permits a one-time adjustment of the antenna resonant frequency, and indicates what type (if any) balun is required. Recalling that standing waves on the antenna feedline results only from impedance mismatch between the antenna and the feedline drives home the significance of the data thus gathered.

282 Standing-wave ratio and related controversies

The Palomar Model RX-100 noise bridge is used in conjunction with the reactance chart (Fig. 8-24). This chart converts the capacitance reading to reactance at 1 MHz. Notice in Fig. 8-25 that readings to the left of the central position of the pointer knob are *capacitive reactance*, whereas readings to the right are *inductive reactance*. Thus, the chart enables these readings to be converted to be either capacitive or inductive reactance at 1 MHz.

Obviously, most measurements will be conducted at frequencies other than 1 MHz (f_1). Nonetheless, the 1-MHz reactances are recorded from the chart. Conversion to reactance at the actual measurement frequency (f_x) can then be computed. If the 1-MHz reactance indicated from the chart is inductive, the inductive reactance at the measurement frequency, f_x, is f_x/f_1 times the chart value, where both frequencies are expressed in megahertz. For example, if the reactance pointer-knob setting is at 51 X_L when a noise null is obtained, the chart shows that this corresponds to an inductive reactance of 1000 Ω at 1 MHz. Assuming a mea-

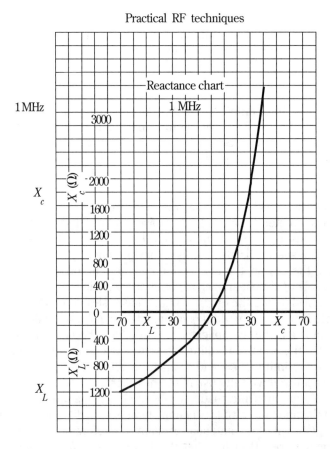

8-24 Reactance chart for Palomar RX-100 R-X noise bridge. These 1-MHz values are readily converted to the values corresponding to the actual measurement frequency.
Palomar Engineers

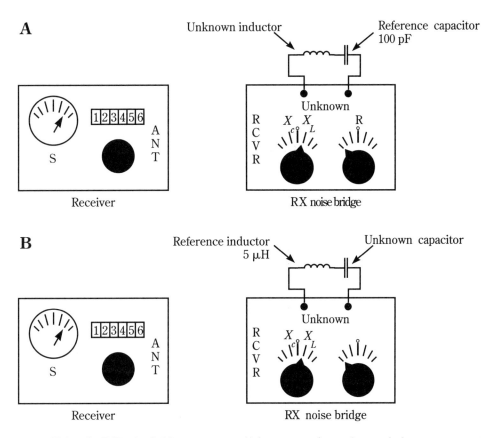

8-25 Using the R-X noise bridge to measure inductance and capacitance. A. An arrangement for measuring unknown inductor. B. An arrangement for measuring unknown capacitor. The objective in both setups is to find the nulling frequency (resonance) when the reactance pointer knob is set at its midpoint position (zero reactance).

surement frequency of, perhaps 5 MHz; the inductive reactance at the measurement frequency is then $5/1 \times 1000 = 5000\ \Omega$.

On the other hand, if the reactance pointer knob indicates capacitive reactance, the reciprocal multiplication is made to convert from the chart indication to capacitive reactance at the measurement frequency. That is, the multiplying factor is now f_1/f_x. For example, suppose that the noise null obtains at the reactance pointer knob at 10 X_C. This time, the chart indicates a capacitive reactance of 400 Ω at 1 MHz. Again, assuming a measurement frequency of 5 MHz, the capacitive reactance at 5 MHz is computed as $\frac{1}{5} \times 400$ or 80 Ω.

Notice that the midway setting of the reactance pointer knob, or zero reactance, can apply to any of five entities being measured:

- A physical resistance with negligible stray reactance.
- An antenna that is self-resonant at the measurement frequency.

- An LCR circuit that can exhibit series or parallel resonance at the measurement frequency.
- A transmission line that is terminated with a resistive load equal to its characteristic impedance. Ideally, this should be valid for any measurement frequency.
- Transmatches, impedance-matching networks, baluns, and transformers operating into their proper resistive loads.

The "R" pointer knob on the noise bridge is always read "as is" no charts or computations are required. However, during any measurement procedure, there is bound to be some interaction between the reactance and resistance adjustments; some "see-sawing" between the two will be required in order to establish the deepest noise null.

An interesting use of the noise bridge is to establish the electrical length of $\frac{1}{2}$-λ and $\frac{1}{4}$-λ transmission lines. For a $\frac{1}{2}$-λ line, the far end should be short-circuited. The noise bridge is then used to monitor the impedance situation at the mending end of the line; at the frequency that corresponds to a $\frac{1}{2} \lambda$ (or any multiple of $\frac{1}{2} \lambda$). The far-end short-circuit should appear at the sending end of the line. This implies a frequency at which the null occurs with both the reactance and resistance pointer knobs set at zero. Thus, either the line can be tailored in length to yield this result at a designated frequency, or you can learn the frequency at which the unmodified line is electrically $\frac{1}{2} \lambda$ (or multiples thereof).

A similar measurement technique can be used for $\frac{1}{4}$-λ determinations. In this case, a short-circuit will appear when the sending-end line is open-circuited. Thus, the idea is to find the noise-nulling frequency at which the noise-bridge resistance and reactance settings are zero. At such a frequency, or odd multiples thereof, $\frac{1}{4}$-λ transmission line-behavior is obtained.

Notice that the $\frac{1}{2}$-λ line is an impedance reproducer, whereas the $\frac{1}{4}$-λ line is an impedance inverter. In either case, you will usually want to determine the lowest test frequency at which the desired line behavior occurs.

From what has been said, an R-X noise bridge that is connected to the sending end of an antenna feedline that is predetermined to be any number of electrical $\frac{1}{2} \lambda$ at a certain frequency can be used to determine the resistance and reactance of the antenna at that frequency. The measurement results should, because of the impedance-reproducing behavior of the line, be essentially the same as if the R-X noise bridge was physically located at the antenna feed point. The practical ramifications of this are that you can use an adjustable line section at the sending end of the line to "scan" the behavior of the antenna over a narrow frequency range.

An excellent use of the R-X noise bridge is to quickly and accurately adjust the antenna tuner or transmatch. Its use for this purpose has the advantage over the traditional deployment of the VSWR indicator in that it is not necessary to put the transmitter on the air. This procedure helps to alleviate some of the interference that plagues the ham bands. Nonetheless, when the measurement procedure is completed, you will find that the VSWR is very close to 1.00 in most instances. The procedure is simple and is carried out as follows:

Connect the UNKNOWN port of the noise bridge to the TRANSMITTER port of the antenna tuner. The transmitter output has been temporarily disconnected from the antenna tuner, and the transmitter remains turned off during the measurement procedure. Connect the RCVR port of the noise bridge to the input of the receiver. The antenna feedline remains connected to the ANTENNA port of the antenna tuner.

With this arrangement, set the noise bridge controls to $X = 0$ (this is the center position of the pointer knob) and $R = 50\ \Omega$. Next, adjust the antenna tuner for a noise null. This simple procedure establishes the antenna-tuner input impedance at 50 Ω resistive, the exact load that the transmitter can deliver maximum power to. Having done this, remove the noise bridge and restore the transmitter, antenna tuner, and receiver to their normal situations. If the nulling procedure has been properly carried out, you will now find that the VSWR in the connecting line between the transmitter and antenna tuner is close to unity. If this is not the case, tune up the transmitter into a 50-Ω dummy load. Also, investigate the possibility of RF on the outside of the coaxial-cable antenna feedline.

In this procedure, the transmitter and receiver have been assumed to be separate for the sake of simplicity. Actually, most amateur stations now use transceivers. This does not change the procedure, but it then becomes important to be aware of precautionary factors. Never turn on the transmitter while the noise bridge is in the measurement arrangement. Otherwise, the noise bridge will probably be damaged as RF enters the bridge arms and the noise amplifier.

The R-X noise bridge can be used to provide a quick and easy test of the integrity of baluns. Ohmmeter tests generally cannot yield this information because both good and bad baluns of the conventional transformer type will exhibit dc short-circuits between input and output. Connect a short coaxial cable between the input of the balun and the UNKNOWN port of the noise bridge. Then, assuming that the impedance ratio of the balun is 1:1, connect a 50-Ω composition resistor to the output of the balun. Connect an AM receiver to the RCVR port of the noise bridge and tune the receiver to the frequency or frequencies of interest. A deep noise null should be attainable with the noise bridge adjusted at $X = 0$ and $R = 50$. Baluns of other impedance ratios are tested in a similar way by appropriately selecting the value of the resistor that is placed across the output winding of the balun.

The RX-100 R-X noise bridge can also provide a fairly precise measurement of inductors and capacitors in sizes that are likely to be encountered in RF systems and circuits through the 1000-kHz to 100-MHz range. For these purposes, you should have a 5-μH reference inductor and a 100-pF mica reference capacitor. The arrangements in Figs. 8-25 AB should be used.

To measure an unknown inductor, the unknown inductor should be connected in series with the reference capacitor and the series combination should be connected to the UNKNOWN port of the noise bridge. The receiver, of course, is connected to the RCVR port of the noise bridge. The receiver is then tuned for a noise null, which should occur at zero X and at roughly the RF resistance of the unknown inductor, which can be determined empirically by searching for the deepest attainable noise null. Record the frequency of the noise null, as indicated by the receiver. Then, use the following relationship to compute the unknown inductance:

$$L = \frac{25{,}330}{f^2 C}$$

where: L is the unknown inductance in microhenries.
C is the reference capacitance in picofarads.

To measure the capacitance of an unknown capacitor, connect it in series with the reference inductor and the series combination, and the series combination should be connected to the UNKNOWN port of the noise bridge. As before, the receiver input is connected to the RCVR port of the noise bridge. The receiver is then tuned for a noise null that should occur at zero X and at roughly the RF resistance of the reference inductor, which can be determined empirically by searching for the deepest attainable noise null. Record the frequency of the noise null, as indicated by the receiver. Then, use the following relationship to compute the unknown capacitance:

$$C = \frac{25{,}330}{f^2 L}$$

where: C is the unknown inductance in picofarads.
L is the reference capacitance in microhenries.

In this measurement procedure, the deepest noise null might occur with the X pointer knob slightly off zero. This could be caused by distributed capacitance in the inductors, or by stray capacitance or stray inductance in the leads. As might be expected, the effects of stray reactance are likely to be most pronounced at higher frequencies. To keep such effects at a minimum, short leads should be used and the components undergoing measurement should be kept as far away from the metal box of the noise bridge as is consistent with keeping the lead lengths short.

Index

A
ac resistance, 242
AM amplifier, 200-203
amplifiers
  1 kW, 230-240
  AM (135-MHz), 200-203
  chain diagrams, 24-26
  class-C, 195-196
  class-C (80-W), 218-221
  class-C tube RF, 7-10
  class-D, 89-92
  class-F, 90
  combining two or more, 17-18
  doubling output power, 202-203
  driver (50-W), 230-240
  driver for SSB, 204-206
  grounded-grid, 13
  JFET broadband linear, 196-197
  linear, 10-13
  linear power MOSFET (2-M), 198-200
  low-power RF modules, 20-23
  marine-band (32-W), 215-216
  mobile 80-W 175-MHz FM, 221-224
  MOSFET broadband linear, 197-198
  output circuits for medium/high-power, 57
  power (330-W), 230-240
  power MOSFET (100-W 80-MHz), 224-227
  quadrature, 212-214
  RF modules, 18-19
  tapered-stripline microwave, 209-212
  test (150-W 28-MHz), 227-230
  wave purity, 108-110
antenna-system tuners, 146-147
applications
  low-frequency, 26-35
  low-power, 191-214
  medium/high power, 215-240
  solid-state, 4, 31

B
balanced transistors, 227
bipolar RF power transistors, 47-73
  class-C amplifier, 52-53
  classes, 69-73
  features, 47-53
  frequency multipliers, 65-67
  gain-leveling provisions, 49-52
  grounding, 65
  matching networks, 53-57
  mounting provisions, 50
  operating regions, 48
  output impedance, 111-112
  problems, 63-67
  reducing hot-spotting, 59-63
  selecting, 67-69
  survival, 58-63
  thermal effects, 63
Bird Thruline RF wattmeters, 186-187
boards (*see* PC boards)
Bruene circuits, 182-185

C
capacitors
  inductance of network, 137-140
  selecting for coupling/bypassing/dc-blocking, 136-137
  using two parallel, 139-140
characteristic impedance, 154-189, 242, 247, 268
class-C amplifiers, 195-196, 218-221
  transistor, 52-53
  tube RF, 7-10
class-D amplifiers, 89-92
class-F amplifiers, 90
classes, 69-73
coaxial cable, measuring velocity, 277-278
common base, 94
common cathode, 94
common collector, 94, 206
common emitter, 94
common grid, 94
common plate, 94
crystal oscillators, 192-193

Index

current standing-wave ratio (CSWR), 241

D
dc resistance, 242
doppler effect, 191
doppler radar, 191-192
dynamic resistance, 242

E
electromagnetic waves, 5

F
field-effect transistors (FET), 75-99
  latest/future releases, 96-98
  operational modes, 76
  power, 79-84
  V-groove gate structure, 81-82
filters
  harmonic, 141-152
  high-pass, 145
  low-pass, 145
FM amplifier, 221-224
frequency multipliers, transistor, 65-67
fundamental frequency, 10

G
grounded-grid amplifier, 13
grounding, common lead, 65

H
harmonic filters, 141-152
  30-MHz, 143-145
hertz (Hz), 1
high-pass filters, 145
high-power applications (see medium/high-power applications)
hot-spotting
  bias considerations, 60
  reducing, 59-63
hybrid transformer, 15

I
impedance
  bipolar transistor output, 111-112
  characteristic, 154-189, 247, 268
  conversions between s and Z parameters, 131
  scattering parameter measurements, 128-132
  transistor, 125-128
impedance-matching networks, 101-152
  capacitors, 136-140
  circuits and semantics, 133-140
  computerized solutions, 115-125
  considerations, 101-102
  design, 102-103
  facts, 110-115
  formulas, 113
  four-element, 150-152
  harmonic filters, 141-152
  high-pass tee, 146-147
  pi-L, 148-150
  SPC transmatch, 147-148
  transformation basics, 103-108
  used with transistor amplifiers, 112-115
  wave purity, 108-110
inductors, toroidal, 39-42
insulated-gate bipolar transistor (IGBT), 98-99
inverters, SCR, 27-31

J
junction field-effect transistors (JFET), 75-79
  1-MHz crystal oscillator, 192-193
  broadband linear amplifier, 196-197
  frequency doubler, 194-195

K
kilohertz (KHz), 1

L
layout diagrams, 23-24
  amplifier chains, 24-26
linear RF amplifier, 10-13
Litz wire, 37-39
  formats, 40
low-pass filters, 145
low-power applications
  AM amplifier, 200-203
  class-C amplifier, 195-196
  common-collector oscillator, 206-209
  crystal oscillator, 192-193
  doubling amplifier output power, 202-203
  JFET broadband linear amplifier, 196-197
  JFET frequency doubler, 194-195
  linear power MOSFET, 198-200
  microwave amplifier, 209-212
  microwave doppler radar system, 191-192
  MOSFET broadband linear amplifier, 197-198
  quadrature amplifier, 212-214

M
marine-band amplifier, 215-216
matching networks, 53-57
  impedance-, 101-152
medium/high-power applications, 215-240
  1 kW amplifier, 230-240
  balanced transistors, 227
  broadband linear amplifier, 216-218
  class-C amplifier, 218-221
  marine-band amplifier, 215-216
  mobile FM amplifier, 221-224
  power MOSFET, 224-227
  test amplifier, 227-230
megahertz (MHz), 2
metal migration, 21
metal-oxide semiconductor field-effect transistors (MOSFET), 79-84
  advantages, 80
  high-power family, 9, 92
  power, 224-227
  power disadvantages, 95
  power driver, 197-20
  VMOS power, 80-88
microwave amplifier, 209-212
microwave doppler radar system, 191-192
monomatch bridge circuits, 179-182

N
negative resistance, 242
networks
  computerized solutions, 115-125
  facts pertaining to all, 110-115
  high-pass tee, 146-147
  impedance-matching, 101-152
  matching, 53-57
  output circuits for medium/high-power amplifiers, 57

parallel-tuned output, 55-57
pi, 107-108
pi-L, 148-150
SPC transmatch, 147-148
tee, 108
nonlinear resistance, 242

O

oscillators
  common-collector, 206-209
  crystal, 192-193

P

PC boards, diagrams, 23-24
positive resistance, 242
power
  calculating/measuring, 42-46
  increasing with solid-state, 13-16
  line loss, 261-262
  loss of, 252-256
  reflected, 257-261
  transferring, 264-286
  transferring over band of frequencies, 256-257
power combiner, 15
  Wilkinson technique, 16-18
power divider, 15
power standing-wave ratio (PSWR), 241
printed-circuit boards (see PC boards)

R

R-X noise bridge, 278-286
radiation, 3-5
radiation resistance, 242
radio-frequency power (see RF power)
reactance, 252, 254
reflectometer measuring devices, 178-187
  Bird Thruline RF wattmeters, 186-187
  Bruene circuits, 182-185
  monomatch bridge circuits, 179-182
  SWR analyzer, 187-189
resistance, 242-243
  types of, 242
RF amplifier modules, 18-19
  low-power, 20-23
  reliability, 20-23
RF amplifier (see amplifiers)

RF power
  bipolar transistors, 47-73
  characteristics, 3
  definition, 1-2
  measuring, 42-46
  radiation factor, 3-5
  solid-state (see solid-state)
  tubes (see tubes)
  uses, 2-3

S

safe-operating area (SOA), 58
SCR inverter, 27-31
  push-pull, 29-31
series-parallel capacitance (SPC) transmatch, 147-148
skin effect, 35-42
  facts, 35-36
  formula, 36-37
solid-state
  advantages, 6-7
  applications, 4, 31
  increasing power, 13-16
  low-frequency transmitters, 31-35
splitter, 15
standing-wave ratio (SWR), 241-249
  analyzer, 187-189
stripline elements, designing, 175-177

T

tank circuits, 53-57
  affects of transistor/tube differences, 56-57
  medium/high-power output, 57
  parallel-tuned output, 55-57
test amplifiers, 227-230
test circuits, 23
thermal fatigue, 21
transformers
  hybrid, 15
  toroidal, 39-42
  transmission-line, 169-175
transistors
  balanced, 227
  bipolar RF power, 47-73
  class-C amplifier, 52-53
  FET, 75-99
  grounded-grid amplifier, 13
  IGBT, 98-99
  impedance, 125-128

JFET, 75-79, 192-197
low-frequency applications, 26-35
measuring RF power, 42-46
MOSFET, 79-84, 92-95, 197-200, 224-227
RF power, 1-5
skin effect, 35-42
survival in RF service, 58-63
VMOS power FET, 80-88
vs. tubes, 1-46
transmission lines, 153-189, 245
  1/2-λ, 166-167
  1/4-λ, 155-159
  1/8-λ, 159-163
  balanced, 245
  characteristic impedance, 154-189, 247, 268
  connections between driver/power amplifiers, 167-175
  determining loss, 261-262
  electrical length, 249-252
  elements, 154
  feeder, 245
  flat, 167-169, 245
  lengths, 164-167, 245
  reflectometer-type instruments, 178-187
  resonant, 245
  stripline elements, 175-177
  transformers, 169-175
  tuned, 169
  velocity factor, 251-252
transmitters, solid-state low-frequency, 31-35
tubes, 5-26
  advantages, 6
  transistors vs., 1-46

V

velocity factor, 251-252
  measuring coaxial cable for, 277-278
virtual resistance, 107
VMOS power FET, 80-88
  features, 82-84
  input circuit, 84-86
  output circuit, 87-88
voltage standing-wave ratio (VSWR), 241, 252-286
  determining line loss, 261-262
  measuring directly, 262-264
  power transfer, 256-257

(VSWR), *cont.*
  power transfer problems, 264-286
  reflected power, 257-261

W

Wilkinson power-combining technique, 16-18

winding techniques, 41
winding toroidal inductors, 39-42
wire, Litz, 37-40

Other Bestsellers of Related Interest

Encyclopedia of Electronic Circuits, Volume 1
—Rudolf F. Graf
Reference tool for the hobbyist, technician, student and design professional that contains 1,300 circuit designs, divided into 100 categories and meticulously indexed.

0-07-157328-3, 0-8306-1938-0 $34.95 Paper
0-07-157322-4, 0-8306-0938-5 $60.00 Hard

Encyclopedia of Electronic Circuits, Volume 2
—Rudolf F. Graf
Volume 2 covers over 700 different circuits than in Volume 1, and is again in the same easy-to-use format.

0-07-155958-2, 0-8306-3138-0 $34.95 Paper
0-07-155949-3, 0-8306-9138-3 $60.00 Hard

Encyclopedia of Electronic Circuits, Volume 3
—Rudolf F. Graf
Volume 3 offers 768 pages of all-new circuit diagrams, organized alphabetically in 121 categories and completely indexed. Each circuit lists its source so readers can find more information if desired.

0-07-155822-5, 0-8306-3348-0 $34.95 Paper
0-07-155814-4, 0-8306-7348-2 $60.00 Hard

Encyclopedia of Electronic Circuits, Volume 4
—Rudolf F. Graf and William Sheets
The fourth volume in this bestselling series includes schematics and descriptions for the latest circuitry used in computers, controls, instrumentation, telecommunications, sensors, and numerous other electronics applications. An ideal companion to previous volumes, this monumental work features more than 1,000 all-new circuits in 100 alphabetically arranged categories.

0-07-157712-2, 0-8306-3895-4 $34.95 Paper
0-07-011042-5, 0-8306-3896-2 $60.00 Hard

Encyclopedia of Electronic Circuits, Volume 5
—Rudolf F. Graf and William Sheets
The most comprehensive, flawlessly organized guide to modern electronic circuits available anywhere. Looks at hundreds of state-of-the-art electronic and integrated circuit designs. Includes cumulative index.

0-07-011077-8 $34.95 Paper
0-07-011076-X $60.00 Hard

Electronic Components: A Complete Reference for Project Builders
—Delton T. Horn

Get the most out of almost any electronic component. This benchtop reference catalogs characteristics, specifications, and component uses that range in complexity from basic wire and solder to transistors and ICs. And it presents insights into the theory and operation of components in typical circuit designs, the pros and cons of using devices in various situations, where and how to find parts, and criteria for making substitutions.

0-07-157660-6 $21.95 Paper

How to Order

Call 1-800-822-8158
24 hours a day,
7 days a week
in U.S. and Canada

Mail this coupon to:
McGraw-Hill, Inc.
P.O. Box 182067
Columbus, OH 43218-2607

Fax your order to:
614-759-3644

EMAIL
70007.1531@COMPUSERVE.COM
COMPUSERVE: GO MH

Shipping and Handling Charges

Order Amount	Within U.S.	Outside U.S.
Less than $15	$3.50	$5.50
$15.00 - $24.99	$4.00	$6.00
$25.00 - $49.99	$5.00	$7.00
$50.00 - $74.49	$6.00	$8.00
$75.00 - and up	$7.00	$9.00

EASY ORDER FORM— SATISFACTION GUARANTEED

Ship to:
Name _____
Address _____
City/State/Zip _____
Daytime Telephone No. _____

Thank you for your order!

ITEM NO.	QUANTITY	AMT.

Method of Payment:
☐ Check or money order enclosed (payable to McGraw-Hill)
☐ Discover ☐ American Express
☐ VISA   ☐ MasterCard

Shipping & Handling charge from chart below	
Subtotal	
Please add applicable state & local sales tax	
TOTAL	

Account No. ☐☐☐☐☐☐☐☐☐☐☐☐☐☐☐

Signature _____ Exp. Date _____
Order invalid without signature

In a hurry? Call 1-800-822-8158 anytime, day or night, or visit your local bookstore.

Key = BC95ZZA